Complete Guide for
Growing Plants Hydroponically

Complete Guide for
Growing Plants
Hydroponically

J. Benton Jones, Jr.

GroSystems, Inc.
Anderson, South Carolina, USA

CRC Press
Taylor & Francis Group
Boca Raton London New York

CRC Press is an imprint of the
Taylor & Francis Group, an **informa** business

CRC Press
Taylor & Francis Group
6000 Broken Sound Parkway NW, Suite 300
Boca Raton, FL 33487-2742

© 2014 by Taylor & Francis Group, LLC
CRC Press is an imprint of Taylor & Francis Group, an Informa business

No claim to original U.S. Government works

Version Date: 20130923

International Standard Book Number-13: 978-1-4398-7668-8 (Paperback)

Library of Congress Cataloging-in-Publication Data

Jones, J. Benton, 1930-
　　Complete guide for growing plants hydroponically / J. Benton Jones, Jr. -- First edition.
　　　　pages cm
　　Includes bibliographical references and index.
　　ISBN 978-1-4398-7668-8 (alk. paper)
　　1. Hydroponics. 2. Horticulture. I. Title.

SB126.5.J645 2014
631.5'85--dc23   2013032564

**Visit the Taylor & Francis Web site at
http://www.taylorandfrancis.com**

**and the CRC Press Web site at
http://www.crcpress.com**

Contents

List of Figures

List of Tables

Preface

This book provides valuable information for the commercial grower, the researcher, the hobbyist, and the student—all those interested in hydroponics and how this method of plant production works as applied to a wide range of growing conditions. Students interested in experimenting with various hydroponic growing systems as well as in how to produce nutrient element deficiencies in plants are given the needed instructions.

The book begins with the concepts of how plants grow and then describes the requirements necessary for success when using various hydroponic growing methods. The major focus is on the nutritional requirements of plants and how best to prepare and use nutrient solutions to satisfy the nutrient element requirement of plants using various growing systems and rooting media under a wide range of environmental conditions. Many nutrient solution formulas are given as well as numerous tables and illustrations. Various hydroponic systems of growing are described, giving their advantages and disadvantages. Included are those procedures required to establish and maintain a healthy rooting environment.

This is the fourth book on hydroponics written by the author. The first book was published in 1983, with revisions published in 1997 and 2005. The two initial editions were primarily devoted to describing various techniques for growing plants without soil. These topics were revised to reflect advances that had been made in understanding how plants grow and the influence that the rooting media and atmospheric environments have on plant performance. In the 2005 edition, two new chapters were added—one on the design and function of a hydroponic greenhouse and the other on hydroponic methods for crop production and management. These two new chapters provided the reader with essential information on greenhouse design and function, giving detailed instructions on how to grow various crops hydroponically in the greenhouse as well as outdoors. Although most hydroponic crops are grown commercially in environmentally controlled greenhouses, hydroponic methods and procedures suited for the hobby grower and techniques for outdoor hydroponics were also included. Organic hydroponics was also a topic included in the 2005 edition.

This book primarily focuses on the basic principles and application requirements for growing plants hydroponically. Most of the available hydroponic texts are outdated, while several of the more current texts contain extraneous material of interest to particular growers and describe the design and operation of growing shelters as well as nonhydroponic growing methods, such as the use of soilless organic rooting media. The reader will find in this text detailed information that relates especially to the growing of plants hydroponically and methods of growing that are applicable to a range of environmental growing systems.

The use of trade names and the mention of a particular product in this book do not imply endorsement of the product named or criticism of similar ones not named. Rather, such a product is used as an example for illustration purposes.

Additional material is available from the CRC website
http://www.crcpress.com/product/isbn/9781439876688

J. Benton Jones, Jr.

About the Author

J. Benton Jones, Jr., has written extensively on hydroponic topics and has been engaged in hydroponic research projects for much of his professional career. After obtaining a BS degree in agricultural science from the University of Illinois, he served in the US Navy for 2 years, which included a brief visit to the hydroponic gardens on the island of Okinawa for the purchase of tomatoes and lettuce. After discharge from active duty, he entered graduate school, obtaining MS and PhD degrees from the Pennsylvania State University in agronomy. For 10 years, Dr. Jones served as research professor at the Ohio Agricultural Research and Development Center (OARDC) at Wooster. During this time, he served on an advisory panel working with the greenhouse tomato growers located in the Cleveland, Ohio area.

Joining the University of Georgia (UGA) faculty in 1968, Dr. Jones served in various research and administrative positions. He was actively engaged in hydroponic research, advising hydroponic growers, giving talks, and writing research papers and technical articles on various aspects of the hydroponic technique. He attended all the Hydroponic Society of America annual meetings, frequently serving as a speaker. He was present at the Hydroponics Worldwide: State of the Art in Soilless Crop Production Conference held in Honolulu, Hawaii, in 1985 when Dr. Allen Cooper and his colleagues presented papers on their newly developed Nutrient Film Technique (NFT) (see Savage 1985). After retiring from UGA, Dr. Jones continued his hydroponic research, frequently spoke at hydroponic conferences, continued to write articles for various magazines, and briefly served as southeastern regional director for CropKing, Inc.

Dr. Jones is an avid vegetable gardener, growing vegetables hydroponically in GroBoxes and GroTroughs that he has developed for use in a home garden setting (www.hydrogrowystems.com).

In 1983, Dr. Jones authored his first book on hydroponics, *A Guide for the Hydroponic and Soilless Culture Grower*. A revised edition titled *Hydroponics: A Practical Guide for the Soilless Grower* was published in 1997 and a second edition appeared in 2005.

Dr. Jones maintains two websites: www.hydrogrosystems.com and www.growtomatoes.com. He has an extensive hydroponic library, books, bulletins, research and technical papers, all editions of the Hydroponic Society of America proceedings, and all issues of *The Growing Edge* magazine.

Dr. Jones is considered an authority on applied plant physiology and the use of analytical methods for assessing the nutrient element status of rooting media and plants as a means for ensuring plant nutrient element sufficiency in both soil and soilless crop production settings. At various times, he has served as a director of several university and commercial soil and plant analysis laboratories, and he still serves as an advisor for two such laboratories.

The author is available to conduct library research and other types of investigative work associated with hydroponics. He can be contacted by mail at GroSystems, Inc., 109 Concord Road, Anderson, SC 29621, and by e-mail at: jbhdro@carol.net

chapter one

Introduction

Introduction

The word hydroponics has its derivation from combining the two Greek words, *hydro*, meaning water, and *ponos*, meaning labor (i.e., working water). The word first appeared in a scientific magazine article (*Science*, 178:1) published in February 1937 and was authored by W. F. Gericke, who had accepted this word as was suggested by Dr. W. A. Setchell at the University of California. Dr. Gericke began experimenting with hydroponic growing techniques in the late 1920s and then published one of the early books on soilless growing (Gericke 1940). Later he suggested that the ability to produce crops would no longer be "chained to the soil but certain commercial crops could be grown in larger quantities without soil in basins containing solutions of plant food." What Dr. Gericke failed to foresee was that hydroponic growing would be essentially confined to enclosed environments for growing high cash value crops and would not find its way into the production of a wide range of commercially grown crops in an open environment.

Hydroponics defined

I went to three dictionaries and three encyclopedias to find how hydroponics is defined. *Webster's New World College Dictionary*, fourth edition, 1999, defines hydroponics as "the science of growing or the production of plants in nutrient-rich solutions or moist inert material, instead of soil"; the *Random House Webster's College Dictionary*, 1999, as "the cultivation of plants by placing the roots in liquid nutrient solutions rather than in soils; soilless growth of plants"; and the *Oxford English Dictionary*, second edition, 1989, as "the process of growing plants without soil, in beds of sand, gravel, or similar supporting material flooded with nutrient solutions."

In the *Encyclopedia Americana*, international edition, 2000, hydroponics is defined as "the practice of growing plants in liquid nutrient cultures rather than in soil"; in the *New Encyclopedia Britannica*, 1997, as "the cultivation of plants in nutrient-enriched water with or without the mechanical support of an inert medium, such as sand or gravel"; and in the *World Book Encyclopedia*, 1996, as "the science of growing plants without soil."

The most common aspect of all these definitions is that hydroponics means growing plants without soil, with the sources of nutrient elements as either a nutrient solution or nutrient-enriched water; an inert mechanical root support (sand or gravel) may or may not be used. It is interesting to note that in only two of the six definitions is hydroponics defined as a "science."

Searching for definitions of hydroponics in various books and articles, the following were found. Devries (2003) defines hydroponic plant culture as "one in which all nutrients are supplied to the plant through the irrigation water, with the growing substrate being soilless (mostly inorganic), and that the plant is grown to produce flowers or fruits that are harvested for sale." In addition, he states,

> Hydroponics used to be considered a system where there was no growing media at all, such as the Nutrient Film Technique in vegetables. But today it's accepted that a soilless growing medium is often used to support the plant root system physically and provide for a favorable buffer of solution around the root system. (Devries 2003)

Resh (1995) defines hydroponics as "the science of growing plants without the use of soil, but by use of an inert medium, such as gravel, sand, peat, vermiculite, pumice, or sawdust, to which is added a nutrient solution containing all the essential elements needed by the plant for its normal growth and development." Wignarajah (1995) defines hydroponics as "the technique of growing plants without soil, in a liquid culture." In an *American Vegetable Grower* article entitled, "Is Hydroponics the Answer?" (Anon. 1978), hydroponics was defined for the purpose of the article as "any method which uses a nutrient solution on vegetable plants, growing with or without artificial soil mediums [*sic*]." Harris (1977) suggested that a modern definition of hydroponics would be "the science of growing plants in a medium, other than soil, using mixtures of the essential plant nutrient elements dissolved in water." Jensen (1997) stated that hydroponics "is a technology for growing plants in nutrient solutions (water containing fertilizers) with or without the use of an artificial medium (sand, gravel, vermiculite, rockwool, perlite, peat moss, coir, or sawdust) to provide mechanical support." In addition, Jensen defined the growing of plants without media as "liquid hydroponics" and with media as "aggregate hydroponics."

Similarly related hydroponic terms are "*aqua* (water) culture," "hydroculture," "nutriculture," "soilless culture," "soilless agriculture," "tank farming," or "chemical culture." A hydroponicist is defined as one who practices hydroponics, and hydroponicum defined as a building or garden in which hydroponics is practiced.

Actually, hydroponics is only one form of soilless culture. It refers to a technique in which plant roots are suspended in either a static, continuously aerated nutrient solution or in a continuous flow or mist of nutrient solution. The growing of plants in an inorganic substance (such as sand, gravel, perlite, or rockwool) or in an organic material (such as sphagnum peat moss, pine bark, or coconut fiber) that are periodically watered with a nutrient solution should be referred to as soilless culture but not necessarily hydroponic. Some may argue with these definitions, as the common conception of hydroponics is that plants are grown without soil, with 16 of the 19 required essential elements (see Chapter 3) provided by means of a nutrient solution (see Chapter 4) that periodically bathes the roots.

Most of the books on hydroponic culture (see references at the end of the book) focus on the general culture of plants and the design of the growing system, giving only sketchy details on the rooting bed design and the composition and management of the nutrient solution. Although the methods of solution delivery and plant support media may vary considerably among hydroponic systems, most have proven to be workable, resulting in reasonably good plant growth.

However, there is a significant difference between a "working system" and one that is commercially viable. Unfortunately, many workable soilless culture systems are not commercially viable. Most books on hydroponics would lead one to believe that hydroponic culture methods for plant growing are relatively free of problems since the rooting media and supply of nutrient elements can be controlled. Jensen (1997), in his overview, stated, "Hydroponic culture is an inherently attractive, often oversimplified technology, which is far easier to promote than to sustain. Unfortunately, failures far outnumber the successes, due to management inexperience or lack of scientific and engineering support." Experience has shown that hydroponic growing requires careful attention to details and good growing skills. Most hydroponic growing systems are not easy to manage by the inexperienced and unskilled. Soil growing is more forgiving of errors made by the grower than are most hydroponic growing systems, particularly those that are purely hydroponic.

Is hydroponics a science?

This question has been frequently asked without a definite answer. Most dictionaries do not define hydroponics as a science, but rather as another means of growing or cultivating plants. However, the *Webster's New World College Dictionary*, fourth edition (1999), does define hydroponics as "the science of growing or the production of plants in a nutrient-rich solution." I would assume that the science aspect is that associated with "in a nutrient-rich solution." Not even in the Wikipedia (www.wikipedia.com) definition and accompanying description of hydroponics does the word "science" appear. Probably the only organization actively engaged in the science aspect is the National Aeronautic Space Administration (NASA) since some form of hydroponics will be the selected method for growing plants in space or on celestial bodies. The *Merriam Webster's Collegiate Dictionary*'s definition for science is "something (as a sport or technique) that may be studied or learned like systematized knowledge."

Hydroponics is indeed a technique for growing plants and there has accumulated a body of knowledge regarding how to grow plants using a hydroponic method (or should it be *the* hydroponic method?), therefore fitting the criterion for being a "science" based on the preceding definition. Also, there is an accumulated body of "systemized knowledge" that fits the second part of the science definition.

Hydroponic terminology

As with every technical subject, there develops a language, as well as a jargon, that becomes accepted by those researching and applying that technology. However, the developed language and/or jargon can be confusing to those unfamiliar with the technology, and sometimes even for those within. Therefore, the hydroponic literature can be confusing to readers due to the variety of words and terms used. The words "hydroponic" and "soilless" grower have and are still being used to refer to the same method of growing, but in this text the word "hydroponic" is used when growing systems are purely hydroponic—that is, there is no rooting medium or the rooting medium is considered inert. The word "soilless" is used for systems of growing that relate to plant production in which the medium can interact with plant roots, such as organic substances such as peat moss and pine bark.

In the organically based developing plant science technology, there are two words that are frequently used: *food* and *nutrient*. It can be confusing if these words are not clearly defined and understood.

What came into common use, beginning in the 1950s, was the word *food* to identify a chemical fertilizer, a substance that contains one or several of the essential plant elements. Today in both agronomic and horticultural literature, it is not uncommon to identify a NPK (nitrogen, phosphorus, potassium) fertilizer as plant food, a word combination that has been generally accepted and commonly used and understood. One dictionary definition of food related to plants is "inorganic substances absorbed by plants in gaseous form or in water solution" (*Merriam Webster's Collegiate Dictionary*, 10th ed., 1994). This dictionary definition would be in agreement with the word combination *plant food*, since chemical fertilizers are inorganic and root absorption of the elements in a chemical fertilizer takes place in a water solution environment. Therefore, the words *food* and/or *plant food* would not relate to organically based substances for use as fertilizer since these two terms have already been defined to identify inorganic substances. Therefore, those organic substances for use as a fertilizer should be identified by name rather than as either a *food* or *plant food*.

The word *nutrient* is vague in its meaning and used in many different scientific fields. A dictionary definition does not help as it is not specific, being defined as "a nutritive substance or ingredient." For plant nutrition application, *nutrient* is understood as being one of the thirteen plant essential mineral elements that have been divided into two categories: the six major mineral elements—N, P, K, Ca, Mg, and S—found at percent concentrations in the plant dry matter, and the seven micronutrients—B, Cl, Cu, Fe, Mn, Mo, and Zn—found in the dry matter of the plant at less than 100% levels (see pp. 35–37). For designating one of the thirteen plant essential-mineral elements, the term *plant nutrient element* is frequently used, such as stating that P is an essential plant nutrient element. Using the term *nutrient element* does not give the proper identification as that being associated with plants.

Unfortunately, the terminology used in both scientific and technical plant journals has been sloppy in identification of the essential plant mineral elements, referring to them as *essential nutrients, plant nutrients,* or just the word *nutrient.* For those engaged in the plant sciences, most generally understand what these terms mean, but for someone not so engaged, the word *nutrient* could have meaning for a wide range of substances as being "a nutritive substance or ingredient."

In the organically based plant growing jargon, the word *nutrient* is used as an all-inclusive term that also includes organic compounds containing combined and bonded carbon, hydrogen, and oxygen. Therefore, one might ask, "What is the difference between a plant mineral element and a substance identified as a nutrient that is an organic substance?" (Parker 1981; Landers 2001). This question is difficult to answer since the criteria for establishing essentiality for the plant mineral elements have been already established (Arnon and Stout 1939; see p. 34), while criteria of essentiality for other than a mineral element have not. Therefore, as with the use of the words *food* or *plant food,* the use of the word *nutrient* should be confined to the identification of only a plant essential mineral element; those suggesting plant nutritive value for an organic substance should use only the word for that substance and not identify it as a nutrient.

Historical past

The growing of plants in plant nutrient element-rich water has been practiced for centuries. For example, the ancient Hanging Gardens of Babylon and the floating gardens of the Aztecs in Mexico were hydroponic in nature. In the 1800s, the basic concepts for the hydroponic growing of plants were established by those investigating how plants grow (Steiner

1985). The soilless culture of plants was then popularized in the 1930s in a series of publications by a California scientist (Gericke 1929, 1937, 1940).

During the Second World War, the US Army established large hydroponic gardens on several islands in the western Pacific to supply fresh vegetables to troops operating in that area (Eastwood 1947). Since the 1980s, the hydroponic technique has come into commercial use for vegetable and flower production, with over 86,000 acres of greenhouse vegetables being grown hydroponically throughout the world—an acreage that is expected to continue to increase (Resh 2013).

One of the aspects of hydroponics that has influenced its protocols is the fact that the hydroponic technique for growing plants is used primarily in controlled environments, such as greenhouses, where the air surrounding plants and its temperature, humidity, and movement are controlled. Even the impact of solar radiation is somewhat controlled (modified) by the transmission characteristics of the greenhouse glazing. Therefore, those reporting on their use of a particular hydroponic method are making observations that are the result of the interaction between the plant environment and growing technique, whether it be a flood-and-drain (see p. 108), NFT (see p. 104), or drip irrigation method (p. 110) with plants rooted in rockwool or coir slabs, or buckets of perlite (see pp. 89–97). Then the question concerns the value of information being reported when plants are grown in a glass greenhouse located in the mountains of Arizona using the rockwool slab drip irrigation system for someone who may be contemplating growing the same plant species in a double polyethylene-covered greenhouse located in the coastal plains of south Georgia (United States) using the rockwool slab drip-irrigation growing method.

Several years ago, I made frequent visits to four hydroponic growers, one located in Georgia and three in South Carolina. All were growing tomatoes in double-layered polyethylene-covered Quonset-type greenhouses (Figure 1.1), with the tomato plants rooted in perlite-filled BATO buckets using a drip irrigation system for delivering a nutrient solution (Figure 1.2). I quickly learned that the skill of these growers was a major factor affecting their obtained yield and fruit quality. All were following the operational procedures provided by the supplier of the greenhouse and hydroponic growing system. Each grower had experienced several instances of plant nutrient element insufficiencies, and as a result, one had made a major change in the nutrient solution formulation he was initially using. All were doing fairly well in terms of fruit yield and quality, although additional experience would probably have resulted in an improvement in both.

In the following growing season things changed: Both fruit yield and quality declined, as all struggled to adjust their operational procedures to cope with what was occurring, but without success. One grower ended

Figure 1.1 Quonset-style double-wall polyethylene-covered greenhouse (common design for most single-bay greenhouses primarily for use by an owner/grower operator).

his crop in midseason; the other three searched for an answer to why things had changed—asking me at each visit as well as making frantic telephone calls to those who had advised them in the past when dealing with other problems. Although no specific cause and effect was uncovered, weather conditions had changed significantly that year: The winter and spring months were warmer than normal, and there were fewer cloudy days with very dry air conditions, with low rainfall leading to drought conditions in the entire area.

Figure 1.2 Drip irrigation line (large black tubing is the delivery line, and the smaller tubing the drip tube) for delivering nutrient solution to perlite-filled BATO buckets.

From weather station data, the minutes of sunshine during this period of time were significantly higher than in previous years. From these data I concluded that the radiation input into their greenhouses was significantly higher than in previous seasons, thereby stressing the plants, with the result being low fruit yield and poor fruit quality (mostly small fruit). What might have helped would have been drawing shade cloth over the plant canopy during the high noon periods of intense radiation. Also needed was a change in the nutrient solution formulation as well as regulating the periods of irrigation in order to reduce the stress that was occurring with the accumulation of "salts" in the perlite. In addition, bringing conditioned air into the greenhouse up through the plant canopy would have kept the plant foliage cooler and contributed to consistent maintenance of plant leaf turgidity. Could these procedures be then considered "systemized knowledge" and, if known and applied, would they have prevented the fruit losses these growers experienced?

From the time that the Hoagland/Arnon nutrient solution formulations were introduced (Hoagland and Arnon 1950; see pp. 60–61), little research has been devoted to investigating the use of these two formulations under various application methodologies. It is not uncommon to read an article in a research journal or technical publication in which the writer uses the term "modified Hoagland nutrient solution" without indicating whether the formulation itself was altered or one of its use parameters. The use parameters given by Hoagland and Arnon were 1 gallon of nutrient solution per plant with replacement each week. Therefore, what effect on plant growth would occur if one of these parameters were changed?—another good question.

From one's real-life experience, the science of hydroponics should be defined based on accompanying environmental conditions; that one set of hydroponic growing procedures would only apply to a particular set of growing parameters and therefore not a fixed set of procedures that would apply universally. Until this is understood, the application of the hydroponic method of growing will flounder in a maze of misinformation; growers will be constantly searching for answers to why things happened as they did without uncovering the cause, and those who want to know the cause will be looking for an answer in all the wrong places.

Proper instruction in the design and workings of a hydroponic culture system is absolutely essential. Those not familiar with the potential hazards associated with these systems or who fail to understand the chemistry of the nutrient solution required for their proper management and plant nutrition will fail to achieve commercial success with most hydroponic culture systems.

The technology associated with hydroponic plant production has changed little as can be seen by reviewing the various bibliographies on hydroponics. Today, those interested in hydroponics seek information from websites on the Internet. The challenge is not lack of information (there are over 400,000 hydroponic websites), but rather the flood of information, much lacking a scientific basis, that leads to confusion and poor decision making on the part of users.

"Is Hydroponics the Answer?" was the title of an article that appeared in 1978 (Anon. 1978) that contained remarks by those prominent at that time in the hydroponic industry. In the article was the following quote: "Hydroponics is curiously slow to receive the mass grower endorsement that some envisioned at one time." Later, Carruthers (1998) provided a possible answer for what had been occurring in the United States, stating that "the reasons for this slow growth can be attributed to many factors, including an abundance of rich, fertile soil and plenty of clean water." At the 1985 Hydroponics Worldwide: State of the Art in Soilless Crop Production conference, Savage (1985) stated in his review that "many extravagant claims have been made for hydroponics/soilless systems, and many promises have been made too soon, but the reality is that a skilled grower can achieve wondrous results." In addition, Savage saw "soilless culture technology as having reached 'adulthood' and rapid maturing to follow," further stating that "soilless and controlled environment crop production takes special skills and training; however, most failures were not the result of the growing method, but can be attributed to poor financial planning, management, and marketing." At the 2003 South Pacific Soilless Culture Conference, Alexander (2003) reported on current developments, stating that "hydroponics is growing rapidly everywhere and within the next 5 to 10 years will be established as a major part of our agricultural and horticultural production industries." His prediction has yet to come true.

Earlier, Wilcox (1980) wrote about the "High Hopes of Hydroponics," stating that "future success in the greenhouse industry will demand least-cost, multiple-cropping production strategies nearer to the major population centers." In 1983, Collins and Jensen (1983) prepared an overview of the hydroponic technique of plant production, and Jensen (1995) discussed probable future hydroponic developments, stating that "the future growth of controlled environment agriculture will depend on the development of production systems that are competitive in terms of costs and returns with open field agriculture," and that "the future of hydroponics appears more positive today than any time over the last 30 years." In a brief review of hydroponic growing activities in Australia, Canada, England, France, and Holland, Brooke (1995) stated that "today's hydroponic farmer can grow crops safely and in places that were formerly considered too barren to cultivate, such as deserts, the Arctic, and even in space." He concluded, "Hydroponic technology spans the globe." Those looking for a brief overview of the common systems of hydroponic growing in use today will find the article by Rorabaugh (1995) helpful.

Naegely (1997) stated that the "greenhouse vegetable business is booming." She concluded, "The next several years promise to be a dynamic time in the greenhouse vegetable industry." Growth in the hydroponic-greenhouse industry was considerable in the 1990s, and its continued future expansion will depend on developments that will keep "controlled environmental agriculture" (CEA) systems financially profitable (see pp. 129). Jensen (1997) remarked that "while hydroponics and climate controlled agriculture (CCA) are not synonymous, CEA usually accompanies hydroponics—their potentials and problems are inextricable."

"Hydroponics for the New Millennium: A Special Section on the Future of the Hydroponic Industry" is the title of a series of articles by six contributors who addressed this topic from their own perspectives; the final comment was, "It really is an exciting time to be in the worldwide hydroponic industry, whether it's for commercial production or a hobby" (*Growing Edge* 11(3):6–13, 2000). Jones and Gibson (2002) stated that "the future of the continued expansion of hydroponics for the commercial production of plants is not encouraging unless a major breakthrough occurs in the way the technique is designed and used." Those factors limiting wide application, wrote Jones and Gibson, "are cost, the requirement for reliable electrical power, inefficiencies in the use of water and nutrient elements, and environmental requirements for disposal of spent nutrient solution and growing media." Schmitz (2004) remarked that "hydroponics is also seen as too technical, too expensive, too everything."

It should be noted that all of these comments regarding hydroponics were written prior to 2000, reflecting the views from many different sources. Since 2000, little has been written about hydroponics in terms of its advantages and disadvantages, and there have been no significant advances that have redirected its application.

The future of hydroponics

Resh (2013) addresses the future of hydroponics by stating that, "in a relatively short period of time, over about 65 years, hydroponics has adapted to many situations from outdoor field culture and indoor greenhouse culture to highly specialized culture in the space program." He also states that the only restraints for the application of hydroponic growing would be "water and nutrients." In addition, Resh sees many future applications of hydroponics in a variety of situations, from food production in desert regions to urban and space applications.

What is not encouraging for the future is the lack of input from scientists in public agricultural colleges and experiment stations that at one time made significant contributions to crop production procedures, including hydroponics. The early hydroponic researchers, Dr. W. F. Gericke and D. R. Hoagland, for example, were faculty members at the University of California, a land-grant university. Today, only a few in similar universities are still active in hydroponic investigations and research. The current status of agricultural cooperative extension programs varies considerably from state to state. In the past, state specialists and county agents played major roles as sources for reliable information, but today these services are being cut back. Also, few of these specialists and agents have any expertise in hydroponics or extensive experience in dealing with greenhouse management issues. Edwards (1999), however, saw a positive role that county extension offices play in providing assistance to those seeking information when he wrote that "the Extension office is often the first place these people contact."

The science of hydroponics is currently little investigated, and much of the current focus is on the application of existing hydroponic techniques. Hydroponics, as a method of growing, is being primarily supported by those in the private sector who have a vested interest in its economic development based on the products that they market.

Another disturbing factor is that the Hydroponic Society of America has not been active since 1997, when it published its last *Proceedings*. The society was founded in 1979 and had been holding annual meetings and publishing proceedings from 1981 through 1997. Also, the International Society of Soilless Culture, an organization that had held meetings and published proceedings in the past, has not been active for several years.

The role that commercial and scientific advancements have on society cannot be ignored when considering what is occurring in hydroponics today. The ease of movement of produce by surface and air transport, for example, allows for growing food products at great distances from their point of consumption. The advent of plastics has had an enormous impact on hydroponics because growing vessels, liquid storage tanks, drip irrigation tubing and fittings, greenhouse glazing materials, and sheeting materials—essential components in all hydroponic/greenhouse operations—are derived from a wide range of plastic materials that vary in their physical and chemical characteristics (Garnaud 1985; Wittwer 1993). The use of computers and computer control of practically every aspect of a hydroponic/greenhouse operation has revolutionized decision making and managerial control procedures. Although one might conclude that hydroponic crop production is becoming more and more a science, there is still much art required that makes this method of plant production a challenge as well as an adventure.

Hydroponic practice and the art of hydroponics

Anyone who wishes to put hydroponics into practice has ready access to all the resources that are needed to be successful and is able to grow plants using one of the various hydroponic growing systems (see Chapter 4) with good results. The challenge is to take those same resources and generate the highest plant yield and quality. Walking into any greenhouse in which plants are being grown hydroponically, I can quickly assess the quality of management skill being applied, the result of applying unique skills that some individuals seem to have—that ability to take a set of operational parameters and make them work effectively and efficiently together. I am one who firmly believes that there are individuals who have what is called a "green thumb," while there are others who do well with the resources they have, but seem to stay at a level of performance below those with a "green thumb." It is similar to those who can cook a delicious meal, while someone else using the same inputs is able to generate a gourmet meal.

Value of the hydroponic method

In 1981, Jensen listed the advantages and disadvantages of the hydroponic technique for crop production, many of which are still applicable today:

- Advantages
  - Crops can be grown where no suitable soil exists or where the soil is contaminated with disease.
  - Labor for tilling, cultivating, fumigating, watering, and other traditional practices is largely eliminated.
  - Maximum yields are possible, making the system economically feasible in high-density and expensive land areas.
  - Conservation of water and nutrients is a feature of all systems. This can lead to a reduction in pollution of land and streams because valuable chemicals need not be lost.
  - Soil-borne plant diseases are more readily eradicated in closed systems, which can be totally flooded with an eradicant.
  - More complete control of the environment is generally a feature of the system (i.e., root environment, timely nutrient feeding, or irrigation), and in greenhouse-type operations, the light, temperature, humidity, and composition of the air can be manipulated.
  - Water carrying high soluble salts may be used if done with extreme care. If the soluble salt concentrations in the water supply are over 500 ppm, an open system of hydroponics may be used if care is given to frequent leaching of the growing medium to reduce the salt accumulations.
  - The amateur horticulturist can adapt a hydroponic system to home and patio-type gardens, even in high-rise buildings. A hydroponic system can be clean, lightweight, and mechanized.
- Disadvantages
  - The original construction cost per acre is great.

- Trained personnel must direct the growing operation. Knowledge of how plants grow and of the principles of nutrition is important.
- Introduced soil-borne diseases and nematodes may be spread quickly to all beds on the same nutrient tank of a closed system.
- Most available plant varieties adapted to controlled growing conditions will require research and development.
- The reaction of the plant to good or poor nutrition is unbelievably fast. The grower must observe the plants every day.

Wignarajah (1995) gave the following advantages of hydroponics over soil growing:

- All of the nutrients supplied are readily available to the plant.
- Lower concentrations of the nutrient can be used.
- The pH of the nutrient solution can be controlled to ensure optimal nutrient uptake.
- There are no losses of nutrients due to leaching.

Wignarajah (1995) gave only one disadvantage of hydroponic systems: "that any decline in the O_2 tension of the nutrient solution can create an anoxic condition which inhibits ion uptake." His recommendation is that only aeroponics solves this problem since it provides a "ready supply of O_2 to the roots, hence never becomes anoxic."

Internet

The role of the Internet has changed and will continue to change how society educates itself. One can obtain the information and devices needed to establish and manage any type of hydroponic growing system off the Internet. But, the Internet is "awash" with innumerable hydroponic websites, and the challenge is how to separate that which is reliable and true from that which is not true or reliable while wading through the mass of material that exists.

Units of measure

The hydroponic literature can be confusing to readers due to the variety of words and terms used as well as a mix of British and metric units. In this book, when required to clarify the text, both British and metric units are given.

Abbreviations

In order to make the reading of the text easier, abbreviations are used for elements, ions, compounds, and units of measure. If a possibility of confusion exists, both the word and its abbreviation will be used. The following are the abbreviations used in this book:

Elements and Their Symbols

Element	Symbol
Aluminum	Al
Boron	B
Calcium	Ca
Chromium	Cr
Chlorine	Cl
Cobalt	Co
Copper	Cu
Hydrogen	H
Iron	Fe
Lead	Pb
Lithium	Li
Magnesium	Mg
Manganese	Mn
Mercury	Hg
Molybdenum	Mo
Oxygen	O
Phosphorus	P
Potassium	K
Silicon	Si
Sodium	Na
Titanium	Ti
Vanadium	V
Zinc	Zn

Compounds	Elemental formula
Ammonia	NH_3
Ammonium nitrate	NH_4NO_3
Ammonium phosphate	$NH_4H_2PO_4$
Ammonium sulfate	$(NH_4)_2SO_4$
Borax	$Na_2B_4O_7 10H_2O$
Boric acid	H_3BO_3
Calcium carbonate	$CaCO_3$
Carbon dioxide	CO_2

Elements and Their Symbols (Continued)

Calcium chloride	$CaCl_24H_2O$
Calcium nitrate	$Ca(NO_3)_2H_2O$
Calcium sulfate	$CaSO_42H_2O$
Copper sulfate	$CuSO_45H_2O$
Diammonium phosphate	$(NH_4)_2PO_4$
Hydrochloric acid	HCl
Ferric sulfate	$Fe_2(SO_4)_3$
Ferrous sulfate	$FeSO_4$
Magnesium carbonate	$MgCO_3$
Magnesium sulfate	$MgSO_4$
Manganese sulfate	$MnSO_4$
Monoammonium phosphate	$NH_4H_2PO_4$
Nitric acid	HNO_3
Phosphoric acid	H_2PO_4
Potassium chloride	KCl
Potassium nitrate	KNO_3
Potassium sulfate	K_2SO_4
Silica	SiO_2
Sodium nitrate	$NaNO_3$
Sulfuric acid	H_2SO_4
Urea	$CO(NH_2)_2$
Zinc sulfate	$ZnSO_4$

Ionic forms	Elemental formula/valance
Aluminum	Al^{3+}
Ammonium	NH_4^+
Borate	BO_3^{3-}
Chloride	Cl^-
Calcium	Ca^{2+}
Copper	Cu^{2+}
Iron (ferrous, ferric)	Fe^{2+} and Fe^{3+}
Magnesium	Mg^{2+}
Manganese	Mn^{2+}
Molybdenum	MoO^{3-}
Phosphate, tri-	PO_4^{3-}
Dihydrogen phosphate	$H_2PO_4^-$
Monohydrogen phosphate	HPO_4^{2-}
Potassium	K^+
Nickel	Ni^{2+}
Nitrate	NO_3^-
Nitrite	NO_2^-

Elements and Their Symbols (Continued)

Silicate	SiO_4^-
Sulfate	SO_4^{2-}
Vanadium	V^{2+}
Zinc	Zn^{2+}

Units of Measure

Unit	Abbreviation
Acre	A
Parts per million	ppm
Liter	L
Milliliter	mL
Millimeter	mm
Meter	M
Decimeter	dm
Centimeter	cm
Gram	g
Kilogram	kg
Pound	lb
Feet	ft
Yard	y

chapter two

How plants grow

Introduction

The ancient thinkers wondered about how plants grow. They concluded that plants obtained nourishment from the soil, calling it a "particular juyce" existent in the soil for use by plants. In the sixteenth century, van Helmont regarded water as the sole nutrient for plants. He came to this conclusion after conducting the following experiment:

> Growing a willow in a large carefully weighed tub of soil, van Helmont observed at the end of the experiment that only 2 ounces of soil was lost during the period of the experiment, while the willow increased in weight from 5 to 169 pounds. Since only water was added to the soil, he concluded that plant growth was produced solely by water.

Later in the sixteenth century, John Woodward grew spearmint in various kinds of water and observed that growth increased with increasing impurity of the water. He concluded that plant growth increased in water that contained increasing amounts of terrestrial matter, because this matter is left behind in the plant as water passes through the plant.

The idea that soil water carried "food" (see p. 5) for plants and that plants "live off the soil" dominated the thinking of the times. It was not until the mid- to late eighteenth century that experimenters began clearly to understand how, indeed, plants grow. At about the same time, the "humus" theory of plant growth was proposed and was widely accepted. The concept postulated that plants obtain carbon (C) and essential nutrients (elements) from soil humus. This was probably the first suggestion of what would today be called the "organic gardening (farming)" concept of plant growth and well-being. Experiments and observations made by many since then have discounted the basic premise of the "humus theory" that plant health comes only from soil humus sources.

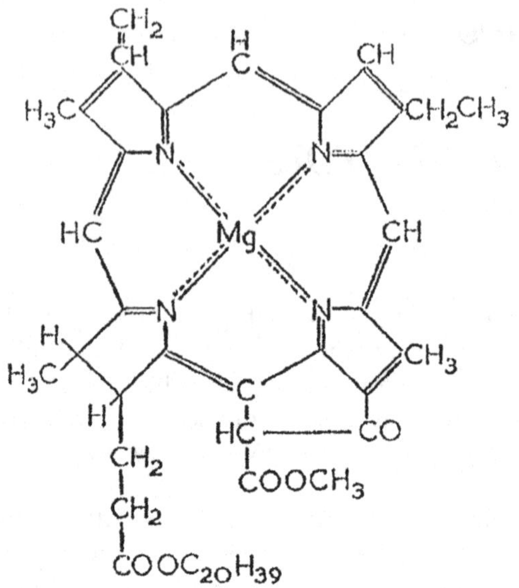

Figure 2.1 Molecular structure of the chlorophyll molecule.

Photosynthesis

Joseph Priestley's famous experiment in 1775 with an animal and a mint plant enclosed in the same vessel established the fact that plants will "purify" rather than deplete the air, as do animals. His results opened a whole new area of investigation. Twenty-five years later, DeSaussure determined that plants consume CO_2 from the air and release O_2 when in the light. Thus, the process that we today call "photosynthesis" was discovered, although it was not well understood by DeSaussure or others at that time.

The process of photosynthesis is the conversion of solar energy into chemical energy in the presence of chlorophyll (Figure 2.1) and light as illustrated in the following formula:

Carbon dioxide ($6CO_2$) + water ($6H_2O$)
in the presence of light and chlorophyll yields
carbohydrate ($C_6H_{12}O_6$) + oxygen ($6O_2$)

A water (H_2O) molecule taken up through the roots is split and then the hydrogen portion is combined with a molecule of CO_2 from the air that has passed into an open stoma to form a carbohydrate, and in the process a molecule of O_2 is released. The rate of photosynthesis is affected by factors external to the plant, such as air temperature (high and low), air movement over the leaf surfaces, level of CO_2 in the air around the leaves, light intensity and its wavelength composition, and water status in the plant. Photosynthesis occurs primarily in green (chlorophyll-containing) leaves, since they have stomata, and not in the other green portions (petioles and stems) of the plant, which do not have stomata. The number of stomata on the leaves and whether they are open or closed will also affect the rate of photosynthesis. Turgid leaves in a continuous flow of air and with open stomata will have the highest photosynthetic rate.

Soil fertility factors

In the middle of the nineteenth century, an experimenter named Boussingault began to observe plants carefully, measuring their growth in different types of treated soil. This was the beginning of many experiments demonstrating that the soil could be manipulated through the addition of manures and other chemicals to affect plant growth and yield. However, these observations did not explain why plants responded to changing soil conditions. Then came a famous report in 1840 by Liebig, who stated that plants obtain all their carbon (C) from CO_2 in the air and the mineral elements by root absorption from the soil. A new era of understanding plants and how they grow emerged. For the first time, it was understood that plants utilize substances in both the soil and the air. Subsequent efforts turned to identifying those substances in soil, or added to soil, that would optimize plant growth in desired directions.

The value and effect of certain chemicals and manures on plant growth took on new meaning. The field experiments conducted by Lawes and Gilbert at Rothamsted (England) led to the concept that substances other than the soil itself can influence plant growth (Rusell, 1950). About this time, the water experiments conducted by Knop and other plant physiologists (a history of how the hydroponic concept was conceived is given by Steiner 1985) showed conclusively that K, Mg, Ca, Fe, and P, along with S, C, N, H, and O, are all necessary for plant life. It is interesting to observe that the formula devised by Knop for growing plants in a nutrient solution can be used successfully today for application in most hydroponic growing systems (Table 2.1).

Table 2.1 Knop's Nutrient Solution Formulation

Reagent	g/L
Potassium nitrate (KNO_3)	0.2
Calcium nitrate [$Ca(NO_3)_2 \cdot 4H_2O$]	0.8
Monopotassium phosphate (KH_2PO_4)	0.2
Magnesium sulfate ($MgSO_4 \cdot 7H_2O$)	0.2
Ferric phosphate ($FePO_4$)	0.1

Keep in mind that the mid-nineteenth century was a time of intense scientific discovery. The investigators named before are but a few of those who made significant discoveries that influenced the thinking and course of scientific biological investigation. Many of the major discoveries of that day centered on biological systems, both plant and animal. Before the turn of the nineteenth century, the scientific basis of plant growth had been well established, as has been reviewed by Russell (1950). Investigators had proven conclusively that plants obtain carbon (C), hydrogen (H), and oxygen (O) required for carbohydrate synthesis from CO_2 and H_2O by the process called photosynthesis (see p. 18); that N was obtained by root absorption of NH_4^+ and/or NO_3^- ions (although leguminous plants can supplement this with symbiotically fixed N_2 from the air); and that all the other elements are taken up by plant roots from the rooting medium as ions and translocated throughout the plant being carried in the transpiration stream.

This general outline remains today as the basis for our present understanding of plant functions. We now know that 16 elements (C, H, O, S, N, P, K, Ca, Mg, B, Cl, Cu, Fe, Mn, Mo, and Zn) are essential for normal plant growth (see p. 31–32). We have extended our knowledge about how these elements function in plants; at what levels they are required to maintain healthy, vigorous growth; and how the elements other than C, H, and O are root absorbed and translocated within the plant.

Although there is much that we do know about plants and how they grow, there is still much that we do not thoroughly understand, particularly about the role of some of the essential elements. Balance, the relationship of one element to another or among the elements, and elemental form may be as important as the concentration of any one of the elements in optimizing the plant's nutritional status. There is still some uncertainty as to how elements are absorbed by plant roots and how they then move within the plant. Elemental form, whether individual ions or complexes, may be as important for movement and utilization as concentration. For example, chelated iron (Fe) forms are effective for control of Fe deficiency, although unchelated ionic Fe, either as ferric (Fe^{3+}) or ferrous (Fe^{2+}) ions, may be equally effective but at higher concentrations.

The biologically active portion of an element in the plant, frequently referred to as the *labile* form, may be that portion of the concentration that determines the character of plant growth. Examples of these labile forms would be the nitrate (NO_3) form of N, the sulfate (SO_4) form of S, and soluble portion of Fe and Ca in plant tissue—forms of these elements that determine their *sufficiency* status. The concept and application of plant analysis (sometimes referred to as tissue testing; see Appendix C) are partly based on this concept of measuring that portion of the element that is found in the plant tissue or its sap, and then relating that concentration to plant growth (Jones 2001, 2012a).

The science of plant nutrition is attracting considerable attention today as plant physiologists determine how plants utilize the essential elements. In addition, the characteristics of plants can now be genetically manipulated by adding and/or removing traits that alter the ability of the plant to withstand biological stress and improve product quality. With these many advances, all forms of growing, whether hydroponic or otherwise, are now becoming more productive. Much of this work is being done for growing plants in space and similarly confined environments where the inputs must be carefully controlled due to limited resources, such as water, and control of the release of water vapor and other volatile compounds into the atmosphere around the plant.

Much of the future of hydroponics may lie with the development of plant cultivars and hybrids that will respond to precise control of the growing environment. The ability of plants to utilize water and the essential elements efficiently may make hydroponic methods superior to what is possible today.

The plant root

Plant roots have two major functions:

- Physically anchor the plant to the growing medium
- Act as an avenue through which water and ions enter into the plant for redistribution to all parts of the plant

Although the first role given here is important, it is the second role that deserves our attention in this discussion. The book edited by Carson (1974) provides detailed information on plant roots and their many important functions, and the book chapter by Wignarajah (1995) discusses the current concepts on nutrient element uptake.

Root architecture is determined by plant species and the physical environment surrounding the roots. Plant roots grow outward and downward. However, in soil, it has been observed that feeder roots grow up, not down. This is why plants, particularly trees, do poorly when the soil surface is compacted or physically disturbed. In soil, any root restriction can have a significant impact on plant growth and development due to the reduction in soil–root contact. Root pruning, whether done purposely (to bonsai plants) or as the result of natural phenomena (due to the presence of plow or clay pans), will also affect plant growth and development in soil.

In most hydroponic growing systems, roots may extend into a much greater volume of growing area or medium than would occur in soil.

Root size, measured in terms of length and extent of branching as well as color, is a characteristic that is affected by the nature of the rooting environment. Normally, vigorous plant growth is associated with long, white, and highly branched roots. It is uncertain whether vigorous top growth is a result of vigorous root growth or vice versa.

Tops tend to grow at the expense of roots, with root growth slowing during fruit set. Shoot-to-root ratios are frequently used to describe the relationship that exists between them, with ratios ranging from as low as 0.5 to a high of 15.0. Root growth is dependent on the supply of carbohydrates from the tops and, in turn, the top is dependent on the root for water and the required essential elements. The loss or restriction of roots can significantly affect top growth. Therefore, it is believed that the goal should be to provide and maintain those conditions that promote good, healthy root development, neither excessive nor restrictive.

The physical characteristics of the root itself play a major role in elemental uptake. The rooting medium and the elements in the medium will determine to a considerable degree root appearance. For example, root hairs will be almost absent on roots exposed to a high concentration (100 mg/L, ppm) of NO_3^-. High P in the rooting medium will also reduce root hair development, whereas changing concentrations of the major cations, K^+, Ca^{2+}, and Mg^{2+}, will have little effect on root hair development. Root hairs markedly increase the surface available for ion absorption and also increase the surface contact between roots and the water film around particles in a soilless medium; therefore, their presence can have a marked effect on water and ion uptake. Normally, hydroponic plant roots do not have root hairs.

The question that arises is, "What constitutes healthy functioning roots for the hydroponic growing system?" The size and extent of root development are not as critical as in soil. It has been demonstrated that one functioning root is sufficient to provide all the essential elements required by the plant, with size and extensiveness of the roots being primarily important for water uptake. Therefore, in most hydroponic systems, root growth and extension are probably far greater than needed, which may actually have a detrimental effect on plant growth and performance. It should be remembered that root growth and function require a continuous supply of carbohydrates, which are generated by photosynthesis. Therefore, an ever expanding and actively functioning root system will take carbohydrates away from vegetative expansion and fruit growth. Therefore, some degree of root growth control may be essential for extensive plant growth and high fruit yields.

A large and extensive root system may not be the best for most hydroponic growing systems. Rather than the large root mass, active, efficiently functioning roots are needed, since the nutrient solution continuously bathes most of the root system, thereby requiring less surface area for absorption to take place. One of the major problems with the NFT (Nutrient Film Technique) tomato hydroponic system (see pp. 104–107), for example, is the large root mass that develops in the rooting channel, which eventually restricts O_2 (Antkowiak 1993) and nutrient solution penetration; the end result is a problem called "root death." Similar extensive root growth occurs with other types of growing systems, particularly with flood-and-drain systems, where roots frequently grow into the piping that delivers and drains the growing bed of nutrient solution, restricting even flow.

Similar extensive root growth is obtained with most hydroponic systems with roots frequently filling bags and blocks of media; in addition, sometimes roots grow through the openings in the outer walls of bags and media containers. The question is, "Does a large root mass translate into high plant performance?" The answer is probably no if there is more root surface for absorption than needed. In addition, roots require a continuous supply of carbohydrates, which can be better used to expand top growth and contribute to fruit yield. A large root mass also requires substantial quantities of O_2 to remain fully functional.

Unfortunately, the question as to root size has yet to be addressed adequately. It should also be remembered that roots require a continuous supply of O_2 to remain healthy and functioning. Roots will not survive in anaerobic conditions. Hydroponically speaking, a large, ever expanding root system probably does not necessarily translate into greater top growth and yield and, in fact, may actually have some detrimental effect.

Water content and uptake

The shape of the plant is determined by its water content, for when the water content declines, wilting occurs and the plant begins to lose its shape and begins to droop. Wilting occurs initially in newly developing tissue that has not yet developed a firm cellular structure. There may be conditions where water uptake and movement within the plant are insufficient to keep the plant fully turgid, particularly when the atmospheric demand is high and/or when the rooting environment (temperature, aeration, and water and salinity levels) is such that it restricts the uptake of water through the roots. In general, field-grown plants are less sensitive to water stress than are plants grown in controlled environments, which may partially explain why plants in the greenhouse are particularly sensitive to water stress, which in turn significantly impacts growth rate and development.

Water is literally pulled up the conductive tissue (mainly in the xylem) by the loss of water from the leaves of the plant by a process called "transpiration," which takes place mainly through open stomata located on leaf surfaces as well as through lenticels and the cuticle. To understand this process, visualize a continuous column of water from the root cells up to atmospherically exposed leaves; the rate of water movement is driven by a water potential gradient between the leaves and the surrounding atmosphere. Transpiration has two important effects: It reduces foliage

Table 2.2 Oxygen Content in Fresh Water Related to
Water Temperature

Temperature		Oxygen content, mg/L (ppm)
°F	°C	
32	0	14.6
41	5	12.8
50	10	11.3
59	15	10.1
68	20	9.1
77	25	8.2
86	30	7.5
95	35	6.9

Source: Nichols, M., 2002, *Growing Edge* 13(5):30–35.

temperature by evaporative cooling (as plant leaves absorb solar energy, most of the absorbed energy is converted into heat), and it provides the physical force for the translocation of elements from the rooting environment up into the upper portions of the plant.

Leaves exposed to direct solar radiation will rise in temperature if water movement up the plant is restricted. Leaf temperature affects rates of photosynthesis, respiration, and plant growth. The amount of water lost by transpiration will depend on the difference in vapor pressure between the leaf and ambient air. Leaf and air temperatures impact gas diffusional rates; hence, rates of photosynthesis and leaf respiration all decrease with increasing leaf temperature. The rate of transpiration increases significantly with increasing movement of air over the leaf surfaces at similar stomata aperture openings. In addition, water lost by transpiration is determined by a complex relationship that exists between air temperature and relative humidity as well as the taxonomic classification and ontogenetic age of the plant organ.

In order for water to enter the roots, the roots must be fully functional. Water absorption by plant roots declines with decreasing temperature, decreases with increasing ion content of the water surrounding the root, and decreases with decreasing O_2 content of the surrounding root mass environment (Table 2.2).

Temperature is another important factor that influences root growth, as well as the absorption of water and essential element ions. The optimum root temperature will vary somewhat with plant species, but in general, root temperatures below 68°F (20°C) begin to bring about changes in root growth and behavior. Below the optimum temperature, there are reduced growth and branching, leading to coarser looking root systems. Absorption of both water and ions is also slowed as the permeability of cell membranes and root kinetics are reduced with decreasing temperature. Translocation in and out of the root is equally slowed at less than optimum root temperatures (68°F to 86°F [20°C to 30°C]). When root temperatures are below the optimum (as well as just being less than the air temperature), plants will wilt during high atmospheric demand periods, and elemental deficiencies will appear. Ion absorption of the elements P, Fe, and Mn seems to be more affected by low temperature than that of most of the other essential elements, major, and micronutrients. It should also be noted that the viscosity of water decreases with decreasing temperature, which in turn affects water movement in and around the plant root.

The maximum root temperature that can be tolerated before significant reduction in root activity occurs is not clearly known. Roots seem to be able to tolerate short periods of high temperature. Roots are fully functional at 86°F (30°C) and probably can withstand temperatures up to 95°°F (35°°C). However, the current literature is not clear as to the exact limits of the optimum temperature range for best plant growth.

In order to avoid the hazards of either low or high temperatures, the roots and rooting medium should be kept at a temperature between 68°F and 86°F (between 20°C and 30°C). Reduced growth and other symptoms of poor nutrition will appear if root temperatures are kept at levels below or above this recommended temperature range.

Aeration is another important factor that influences root and plant growth. Oxygen (O_2) is essential for cell growth and function. If not available in the rooting medium, severe plant injury or death will occur. The energy required for root growth and ion absorption is derived by the process called "respiration," which requires O_2. Without adequate O_2 to support respiration, water and ion absorption cease and roots die.

Oxygen levels and pore space distribution in the rooting medium will also affect the development of root hairs. Aerobic conditions, with equal distributions of water- and air-occupied pore spaces, promote root growth, including root hair development.

If air exchange between the medium and surrounding atmosphere is impaired by overwatering, or the pore space is reduced by compaction, the O_2 supply is limited and root growth and function will be adversely affected. As a general rule, if the pore space of a solid medium, such as soil, sand, gravel, or an organic mix containing peat moss or pine bark, is equally occupied by water and air, sufficient O_2 will be present for normal root growth and function.

In hydroponic systems where plant roots are growing in a standing solution or a flow of nutrient solution, the grower is faced with a "Catch-22" problem in periods of high temperature. The solubility of O_2 in water is quite low (at 75°F, about 0.004%) and decreases significantly with increasing temperature, as is illustrated in Figure 2.2. However, since plant respiration, and therefore O_2 demand, increase rapidly with increasing

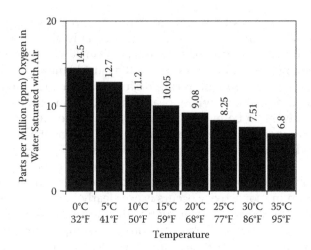

Figure 2.2 Dissolved oxygen (O₂) saturation limits for water at sea level pressure and temperature. (Source: Brooke, L. L., 1995. *Growing Edge* 6 (4): 34–39, 70–71).

temperature, attention to O_2 supply is required. Therefore, the nutrient solution must be kept well aerated by either bubbling air or O_2 into the solution or by exposing as much of the surface of the solution as possible to air by agitation. One of the significant advantages of the aeroponic system (see pp. 108) is that plant roots are essentially growing in air and therefore are being adequately supplied with O_2 at all times. Root death, a common problem in most NFT systems (see pp. 104–107) and possibly other growing systems as well, is due in part to lack of adequate aeration within the root mass in the rooting channel.

In soil and soilless rooting media, a greater root mass can contribute to increasing absorption capacity, while in a hydroponic growing system, root mass is less a contributing factor. The nutritional status of a plant can be a factor, as a healthy actively growing plant will supply the needed carbohydrates required to sustain the roots in an active respiratory condition.

It is generally believed that most of the water absorption by plant roots occurs in younger tissue just behind the root tip. Water movement across the root cortex occurs primarily intercellularly, but can also occur extracellularly with increasing transpiration rate.

As water is pulled into the plant roots, those substances dissolved in the water will also be brought into the plant, although a highly selective system regulates which ions are carried in and which are kept out. Therefore, as the amount of water absorbed through plant roots increases, the amount of ions taken into the root will also increase, even though a regulation system exists. This partially explains why the elemental content of the plant can vary depending on the rate of water uptake. Therefore, atmospheric demand can be a factor affecting the elemental content of the plant, which can be either beneficial or detrimental. In addition, many other water-soluble compounds in the rooting medium might be brought into the plant and enter the xylem.

Ion uptake

All essential mineral element ions in plant root cells are at a higher concentration than that present in the surrounding environment. Therefore, how are the mineral element ions able to move against this concentration gradient? In response, Jacoby (1995) poses the following questions:

1. How is passage through the impermeable liquid layer accomplished?
2. How is accumulation against the concentration gradient accomplished?
3. How is metabolic energy coupled to such transport?
4. What is the mechanism of selectivity?
5. How is vectorial transport accomplished?

The answers to most of these questions have yet to be answered fully. However, the concepts of ion absorption and movement up the plant are described by six processes:

1. Free space and osmotic volume
2. Metabolic transport
3. Transport proteins
4. Charge balance and stoichiometry
5. Transport proteins
6. Transport to the shoot

The absorption of ions by the root is by both a passive and an active process. Depending on the specific ion, transport is by passive uniport through channels or by carrier-aided cotransport with protons (Jacoby 1995). Passive root absorption means that an ion is carried into the root by the passage of water; that is, it is sort of "carried" along in the water taken into the plant. It is believed that the passive mode of transport explains the high concentrations of some ions, such as K^+, NO_3^-, and Cl^-, found in the stems and leaves of some plants. The controlling factors in passive absorption are the amount of water moving into the plant (which varies with atmospheric demand), the concentration of these ions in the water surrounding the plant roots, and the size of the root system. Passive absorption is not the whole story, however, as a process involving chemical selectivity occurs when an ion-bearing solution reaches the root surface.

The membranes of the root cells form an effective barrier to the passage of most ions into the root. Water may move into these cells, but the ions in the water will be left behind in the water surrounding the root. Also, another phenomenon is at work: Ions will only move physically from an area of high concentration to one of lower concentration, a process known as *diffusion*. However, in the case of root cells, the concentration of most ions in the root is normally higher than that in the water surrounding the root. Therefore, ions should move from the root into the surrounding water and, indeed, this can and does occur. The question is how ions move against this concentration gradient and enter the root. The answer is by the process called "active absorption."

In a typical plant root, solutes can be found in three compartments. The outermost compartment, where solutes have ready accessibility, is called apparent free space (AFS) or outer free space (OFS). This compartment contains two subcompartments: water free space (WFS), which dissolved substances (such as ions) can freely move into by diffusion, and Donnan free space (DFS), whose cell walls and membranes have a number of immobile negatively charged sites that can bind cations. The cation exchange capacity of plant cells is determined by the DFS. Ion movement across these cell walls and membranes requires both energy and a carrier system.

Active absorption works based on carriers and Michaelis–Menten kinetics. These theories are based on the nature of cell membranes, which function in several ways to control the flow of ions from outside to inside the cell. It is common to talk about "transporting" an ion across the cell membrane and, indeed, this may be what happens. An ion may be complexed with a particular substance (probably a protein) and then "carried" across (or through) the membrane into the cell against the concentration gradient. For the system to work, a carrier must be present and energy expended. As yet, no one has been able to determine the exact nature of the carrier or carriers, although the carriers are thought to be proteins. However, the carrier concept helps to explain what is observed in the movement of ions into root cells. The other theory relates to the existence and function of ion or proton pumps rather than specific carriers. For both of these systems to work, energy is required: one linked to respiratory energy and the other from adenosine triphosphate (ATP), a high-energy intermediate associated with most energy-requiring processes. For a more detailed explanation on the mechanism of ion uptake by roots, refer to the article by Wignarajah (1995).

Although we do not know the entire explanation for active absorption, general agreement exists that some type of active system regulates the movement of ions into the plant root.

We know three things about ion absorption by roots:

1. The plant is able to take up ions selectively even though the outside concentration and ratio of elements may be quite different from those in the plant.
2. Accumulation of ions by the root occurs across a considerable concentration gradient.
3. The absorption of ions by the root requires energy that is generated by cell metabolism.

A unique feature of the active system of ion absorption by plant roots is that it exhibits ion competition, antagonism, and synergism. The competitive effects restrict the absorption of some ions in favor of others. Examples of enhanced uptake relationships include:

- Potassium (K^+) uptake is favored over calcium (Ca^{2+}) and magnesium (Mg^{2+}) uptake.
- Chloride (Cl^-), sulfate (SO_4^{2-}), and phosphate ($H_2PO_4^-$) uptake is stimulated when nitrate (NO_3^-) uptake is strongly depressed.

The rate of absorption is also different for various ions. The monovalent ions (i.e., K^+, Cl^-, NO_3^-) are more readily absorbed by roots than the divalent (Ca^{2+}, Mg^{2+}, SO_4^{2-}) ions.

The uptake of certain ions is also enhanced in active uptake. If the NO_3^- anion is the major N source in the surrounding rooting environment, then there tends to be a balancing effect marked by greater intake of the cations K^+, Ca^{2+}, and Mg^{2+}. If the NH_4^+ cation is the major source of N, then uptake of the cations K^+, Ca^{2+}, and Mg^{2+} is reduced. In addition, the presence of NH_4^+ enhances NO_3^- uptake. If Cl^- ions are present in sizable concentrations, NO_3^- uptake is reduced.

These effects of ion competition, antagonism, and synergism are of considerable importance to the hydroponic grower in order to avoid the hazard of creating elemental imbalances in the nutrient solution that will, in turn, affect plant growth and fruit development and yield. Therefore, the nutrient solution must be properly and carefully balanced initially and then kept in balance during its term of use. Imbalances arising from these ion effects will affect plant growth. Steiner (1980) has discussed in considerable detail his concepts of ion balance when constituting a nutrient solution.

Unfortunately, many current systems of nutrient solution management do not effectively deal with the problem of imbalance. This is true not only of systems in which the nutrient solution is managed on the basis of weekly dumping and reconstitution but also of constant-flow systems. Indeed, the concept of rapid, constant-flow, low-concentration nutrient solution management is made to look deceptively promising in minimizing the interacting effects of ions in the nutrient solution on absorption and plant nutrition.

Finally, non-ionic substances—mainly molecules dissolved in the water solution surrounding the plant root—can also be taken into the root by mass flow. Substances such as amino acids, simple proteins, carbohydrates, and urea can enter the plant and possibly contribute to its growth and development, something that has not been well documented.

Metabolic transport across root structures into the xylem vessels regulates the number of ions conveyed to the tops; interestingly, the number is little affected by the velocity of xylem sap flow. Once in the xylem, ions and other soluble solutes move by mass flow, primarily to the leaf apoplast.

Root surface chemistry

Many plant roots have the ability to alter the environment immediately around their roots. The most common alteration is a reduction in pH by the emission of hydrogen (H^+) ions. In addition, some plants have the ability to emit substances (such as siderophores) from their roots that enhance ion chelation and uptake. These phenomena have been most commonly observed in species that have the ability to obtain needed Fe under adverse conditions and are characteristic of so-called "iron (Fe)-efficient" plants (Rodriguez de Cianzio 1991).

This ability of roots to alter their immediate environment may be hampered in hydroponic systems where the pH of the nutrient solution is being constantly adjusted upward or in those systems where the nutrient solution is not recycled. In such cases, care must be taken to ensure that the proper balance and supply of the essential elements are provided, since the plant roots may not be able to adjust the rooting environment to suit a particular need.

The impact of roots on a standing aerated nutrient solution system (see pp. 99–103) may have an adverse effect on plant growth by either raising or lowering the solution pH, as well as by the introduction of complexing substances into the solution. Therefore, frequent monitoring of the nutrient solution and close observation of plant growth and development can alert the grower to the nutrient solution's changing status.

chapter three

The essential plant nutrient elements

Introduction

Through the years, a set of terms has been developed to classify those elements essential for plant growth. This terminology can be confusing and misleading to those unfamiliar with it. Even the experienced can become rattled from time to time.

As with any body of knowledge, an accepted jargon develops that is understood well only by those actively engaged in the field. One of the commonly misused terms is referring to the essential metallic elements, such as Cu, Fe, and Zn, as being classed as minerals. The strict definition of mineral refers to a compound of elements rather than a single element. Yet, mineral nutrient is a commonly used term when referring to plant elemental nutrition. This phrase occasionally appears in conjunction with other words, such as plant mineral nutrition, mineral nutrition, or plant nutrition—all of which refer to the essential elements and their requirements by plants.

Another commonly misused and misunderstood word is "nutrient," referring again to an essential element. It is becoming increasingly common to combine the words nutrient and element to mean an essential element. Therefore, elements such as N, P, and K are called "nutrient elements." Unfortunately, no one has suggested an appropriate terminology when talking about the essential elements; thus, the literature on plant nutrition contains a mixture of these terms. In this book, "essential plant nutrient element" is the term used in place of nutrient element or nutrient.

Terminology

The early plant investigators developed a set of terms to classify the 16 elements identified as essential for plants; these terms have undergone changes in recent times. Initially, the major elements—so named because they are found in sizable quantities in plant tissues—included the elements C, H, N, O, P, and K. Unfortunately, three of the now named essential major elements—Ca, Mg, and S—were initially named "secondary" elements. These so-called secondary elements should be classed as major elements, and they are referred to as such in this text.

Those elements found in smaller quantities at first were called "minor elements" or sometimes "trace elements" (B, Cl, Cu, Fe, Mn, Mo, and Zn). More recently, these elements have been renamed "micronutrients," a term that better fits the comparative ratios between the major elements found in sizable concentrations and the micronutrients found at lower concentrations in plant tissues. Another term that has been used to designate some of the micronutrients is "heavy metals," which refers to those elements that have relatively high atomic weights. One definition is "those metals that have a density greater than 50 mg/cm^3," with elements such as Cd, Co, Cu, Fe, Pb, Mo, Ni, and Zn being considered as heavy metals.

Another category that has begun to make its way into the plant nutrition literature is the so-called "beneficial elements," which will be discussed later (see p. 39).

The average concentration of the essential elements in plants is given in Table 3.1, using data by Epstein (1972). More recently, Ames and Johnson (1986) listed the major elements by their internal concentrations found in higher plants, as shown in Table 3.2.

Another recently named category is *trace elements*, which refers to those elements found in plants at very low levels (<1 ppm) but not identified as

Table 3.1 Average Concentrations of Mineral Nutrients Sufficient for Adequate Growth in Plant Dry Matter

Element	Symbol	Dry weight (μmol/g)	mg/kg (ppm)	%	Relative number of atoms
Molybdenum	Mo	0.001	0.1	—	1
Copper	Cu	0.10	6	—	100
Zinc	Zn	0.30	20	—	300
Manganese	Mn	1.0	50	—	1000
Iron	Fe	2.0	100	—	2000
Boron	B	2.0	20	—	2000
Chlorine	Cl	3.0	100	—	3000
Sulfur	S	30	—	0.1	30,000
Phosphorus	P	60	—	0.2	60,000
Magnesium	M	80	—	0.2	80,000
Calcium	Ca	125	—	0.5	125,000
Potassium	K	250	—	1.0	250,000
Nitrogen	N	1000	—	1.5	1,000,000

Source: Epstein, E., 1972, *Mineral Nutrition of Plants: Principles and Perspectives,* John Wiley & Sons, New York.

Table 3.2 Internal Concentrations of Essential Elements in Higher Plants

Element	Symbol	Concentration in dry tissue ppm	%
Major elements			
Carbon	C	450,000	45
Oxygen	O	450,000	45
Hydrogen	H	60,000	6
Nitrogen	N	15,000	1.5
Potassium	K	10,000	1.0
Calcium	Ca	5,000	0.5
Magnesium	Mg	2,000	0.2
Phosphorus	P	2,000	0.2
Sulfur	S	1,000	0.1
Micronutrients			
Chlorine	Cl	100	0.01
Iron	Fe	100	0.01
Manganese	Mn	50	0.005
Boron	B	20	0.002
Zinc	Zn	20	0.002
Copper	Cu	6	0.0006
Molybdenum	Mo	0.1	0.00001

Source: Ames, M. and Johnson, W.S., 1986, in *Proceedings of the 7th Annual Conference on Hydroponics: The Evolving Art, the Evolving Science,* Hydroponic Society of America, Concord, CA.

either essential or beneficial. Some of these trace elements are found in the A–Z micronutrient solution (Table 3.3).

The word "available" has a specific connotation in plant nutrition parlance. It refers to that form of an element that can be absorbed by plant roots. Although its use has been more closely allied with soil growing, it has inappropriately appeared in the hydroponic literature. In order for an element to be taken into the plant, it must be in a soluble form in the water solution surrounding the roots. The available form for most elements in solution is as an ion. It should be pointed out, however, that some molecular forms of the elements can also be absorbed. For example, the molecule urea, $CO(NH_2)_2$ (a soluble form of N); the boric acid molecule, H_3BO_3; and some chelated complexes, such as FeDTPA, can be absorbed by the plant root.

Table 3.3 Elemental Composition of the A–Z Nutrient Solution

A		B	
Reagent	g/L	Reagent	g/L
$Al_2(SO_4)_8$	0.055	As_2O_3	0.0055
KI	0.027	$BaCl_2$	0.027
KBr	0.027	$CdCl_2$	0.0055
TiO_2	0.055	$Bi(NO_3)_2$	0.0055
$SnCl_2 2H^2O$	0.027	Rb_2SO_4	0.0055
LiCl	0.027	K_2CrO_4	0.027
$MnCl_2 4H_2O$	0.38	KF	0.0035
H_3BO_3	0.61	$PbCl_2$	0.0055
$ZnSO_4 7H_2O$	0.055	$HgCl_2$	0.0055
$CuSO_4 3H_2O$	0.055	MoO_3	0.023
$NiSO_4 6H_2O$	0.055	H_2SeO_4	0.0055
$Co(NO_3)_2 6H_2O$	0.055	$SrSO_4$	0.027
		VCl_3	0.0055

Criteria for essentiality

The criteria for essentiality were established by two University of California plant physiologists in a paper published in 1939. Arnon and Stout (1939) described three requirements that an element had to meet in order to be considered essential for plants:

- Omission of the element in question must result in abnormal growth, failure to complete the life cycle, or premature death of the plant.
- The element must be specific and not replaceable by another.
- The element must exert its effect directly on growth or metabolism rather than by some indirect effect, such as by antagonizing another element present at a toxic level.

Some plant physiologists feel that the criteria established by Arnon and Stout may have inadvertently fixed the number of essential elements at the current 16, and that for the foreseeable future no additional elements will be found that meet their criteria for essentiality. The 16 essential elements, the discoverer of each element, the discoverer of essentiality, and the date of discovery are given in Table 3.4; the 16 essential elements, the form utilized by plants, and their function in plants are given in Table 3.5.

Some plant physiologists feel that it is only a matter of time before the essentiality of Co, Ni, Si, and V, known today as *beneficial elements,* will be demonstrated, that those elements should be added to the current list of 16

Table 3.4 Discoverers of Elements and Essentiality for Essential Elements

Element	Discoverer	Year	Discoverer of essentiality	Year
C	a	a	DeSaussure	1804
H	Cavendish	1766	DeSaussure	1804
O	Priestley	1774	DeSaussure	1804
N	Rutherford	1772	DeSaussure	1804
P	Brand	1772	Ville	1860
S	a	a	von Sachs, Knop	1865
K	Davy	1807	von Sachs, Knop	1860
Ca	Davy	1807	von Sachs, Knop	1860
Mg	Davy	1808	von Sachs, Knop	1860
Fe	a	a	von Sachs, Knop	1860
Mn	Scheele	1774	McHargue	1922
Cu	a	a	Sommer	1931
			Lipman and MacKinnon	1931
Zn	a	a	Sommer and Lipman	1926
Mo	Hzelm	1782	Arnon and Stout	1939
B	Gay Lussac and Thenard	1808	Sommer and Lipman	1926
Cl	Scheele	1774	Stout	1954

Source: Glass, D. M., 1989, *Plant Nutrition: An Introduction to Current Concepts,* Jones and Bartlett Publishers, Boston.

[a] Element known since ancient times.

essential plant nutrient elements, and that they should be present in the rooting medium or be added to the rooting medium to ensure best plant growth.

The major elements

Nine of the 16 essential elements are classified as major elements: C, H, O, N, P, K, Ca, Mg, and S. The first three are obtained from CO_2 in the air and H_2O from the rooting medium and then combined by photosynthesis to form carbohydrates via the reaction carbon dioxide (CO_2) + water (H_2O) = (in the presence of light and chlorophyll) carbohydrate ($C_6H_2O_6$) + oxygen ($6O_2$). Thus, they are not normally discussed in any detail as unique to hydroponic growing systems (see p. 18).

The elements C, H, and O represent about 90% to 95% of the dry weight of plants and are indeed the major constituents. The remaining six major elements, N, P, K, Ca, Mg, and S, are more important to the hydroponic grower since these elements must be present in the nutrient solution in sufficient concentration and in the proper balance to meet the crop requirement. Most of the remaining 5% to 10% of the dry weight of plants

Table 3.5 Essential Elements for Plants by Form Utilized and Biochemical Function

Essential elements	Form utilized	Biochemical function in plants
C, H, O, N, S	In the forms of CO_2, HCO_3^-, H_2O, O_2, NO_3^-, NH_4^+, N_2, SO_4^{2-}, SO_2; the ions from the soil solution, the gases from the atmosphere	Major constituents of organic material; essential elements of atomic groups involved in enzymatic processes; assimilation by oxidation-reduction reactions
P, B	In the form of phosphates, boric acid, or borate from the soil solution	Esterification with native alcohol groups of plants; the phosphate esters are involved in energy transfer reactions
K, Mg, Ca, Mn, Cl	In the form of ions from the soil solution	Nonspecific functions establishing osmotic potentials; more specific reactions in which the ion brings about optimum conformation of an enzyme protein (enzyme activation); bridging of the reaction partners; balancing anions; controlling membrane permeability and electropotentials
Fe, Cu, Zn, Mo	In the form of ions or chelates from the soil solution	Present predominantly in a chelated form incorporated in a prosthetic group; enable electron transport by valency charge

Source: Mengel, K. and Kirkby, E. A., 1987, *Principles of Plant Nutrition,* 4th ed., International Potash Institute, Worblaufen-Bern, Switzerland.

is made up of these six elements. A summarization of the important aspects and characteristics of the major elements is found in Appendix B.

The micronutrients

Plants require considerably smaller concentrations of the micronutrients than the major elements to sustain plant nutrient element sufficiency. However, the micronutrients are as critically essential as the major elements are. The optimum concentrations for the micronutrients are typically in the range of 1/10,000 of the concentration range required for the major elements (see Tables 3.1 and 3.2). The micronutrients, as a group, are far more critical in terms of their control and management than some of the major elements. In the case of several of the micronutrients, the required range is quite narrow. Departure from this narrow range results in either deficiency or toxicity when below or above, respectively, the desired concentration range. Deficiency or toxicity symptoms are usually difficult to evaluate visually and therefore require an analysis of the plant for confirmation (see Appendix C).

A deficiency of a micronutrient can usually be corrected easily and quickly, but when dealing with excesses or toxicities, correction can be difficult, if not impossible. If toxicity occurs, the grower may well have to start over. Therefore, great care must be taken to ensure that an excess concentration of a micronutrient not be introduced into the rooting media, either initially or during the growing season.

The availability of some of the micronutrients, particularly Fe, Mn, and Zn, can be significantly changed with a change in pH or with a change in the concentration of one of the major elements, particularly P. Therefore, proper control of the pH and concentration of the major elements in a nutrient solution is equally critical.

There may be sufficient concentration of some of the micronutrients in the natural environment (i.e., in the water used to make a nutrient solution, the inorganic or organic rooting media, or from contact with piping, storage tanks, etc.) to preclude the requirement to supply a micronutrient by addition. Therefore, it is best to analyze a prepared nutrient solution after constituting it and after contact with its environment to determine its micronutrient content. In addition, careful monitoring of the rooting media and plants will ensure that the micronutrient requirement is being satisfied but not exceeded. A summarization of the important aspects and characteristics of the micronutrients is found in Appendix B.

Content in plants

The major elements, C, H, O, N, P, K, Ca, Mg, and S, exist at percentage concentrations in the plant dry matter, while the micronutrients exist at concentrations of 0.01% or less in the dry matter. To avoid confusing decimals, micronutrient concentrations are expressed in milligrams per kilogram (mg/kg) or parts per million (ppm). Other terms are used to define elemental concentration, but % and mg/kg will be the terms used in this book. A comparison of commonly used concentration units is given in Table 3.6.

Plant elemental content varies considerably with species, plant part, and stage of growth—plus the effect of level of elemental availability. Elemental content data are mostly based on dry weight determinations, although some element determinations, such as N (as NO_3^- anion) and P (as $H_2PO_4^-$ and HPO_4^{2-} anions), can also be made using sap extracted from live tissue. An element may not be evenly distributed among various plant parts (roots, stems, petioles, and leaves), and uneven distributions

Table 3.6 Comparison of Commonly Used Concentration Units for Major
Elements and Micronutrients in Dry Matter of Plant Tissue

Elements	Concentration units			
	%	g/kg	cmol(p+)/kg	cmol/kg
Major elements				
Phosphorus (P)	0.32[a]	3.2	—	10
Potassium (K)	1.95	19.5	50	50
Calcium (Ca)	2.00	20.0	25	50
Magnesium (Mg)	0.48	4.8	10	20
Sulfur (S)	0.32	3.2	—	10
Micronutrients				
Boron (B)	20	20	1.85	
Chlorine (Cl)	100	100	2.82	
Iron (Fe)	111	111	1.98	
Manganese (Mn)	55	55	1.00	
Molybdenum (Mo)	1	1	0.01	
Zinc (Zn)	33	33	0.50	

[a] Concentration levels were selected for illustrative purposes only.

may also exist within a leaf and among leaves at varying stages of development. Knowing the concentration of an element in a specific plant part at the known stage of growth, and even its distribution within the plant itself, can provide valuable information for defining the nutritional status of the plant (Jones 2012a).

Function in plants

The primary and secondary roles of all the essential elements required by plants are fairly well known (see Table 3.5). Some elements are constituents of plant compounds (such as N and S, which are constituents of proteins); some serve as enzyme activators (K, Mg, Cu, Mo, Zn), have involvement in energy transfer reactions (P and Fe), are directly and/or indirectly related to photosynthesis (Mg, P, and Fe), or serve as osmotic balancers (K). Some elements have essentially one role or function; others have multiple roles and functions.

Forms of utilization

For all of the essential elements, the form or forms of the element utilized are normally specific as a single ionic form, such as K^+, Ca^{2+}, Mg^{2+}, Cu^{2+}, Mn^{2+}, Zn^{2+}, Cl^-, and Mo_4^{2-}; more than one ionic form, Fe^{2+} and Fe^{3+}; as two-element ions, NO_3^-, NH_4^+, and SO_4^{2-}, as multiple-element ions, HPO_4^{2-} and $H_2PO_4^-$; in two different ionic forms, NO_3^- anion and NH_4^+ cation; or as molecules, H_3BO_3, $CO(CH)_2$ (urea), and silicic acid (H_4SiO_4).

The beneficial elements

The number of elements presently considered essential for the proper nutrition of higher plants stands at 16; the last element added to that list was Cl in 1954 (see Table 3.4). Some plant physiologists feel that the criteria for essentiality established by Arnon and Stout (1939, see p. 34) could preclude the addition of other elements, as these 16 include most of the elements found in substantial quantities in plants. However, there may be other elements that have yet to be proven essential, as their requirements are at such low levels that it will take considerable sophisticated analytical skills to uncover them, or their ubiquitous presence will require special skills to remove them from the rooting medium in order to create a deficiency. This was the case for Cl, the last of the essential elements to be so defined. The question is which elements are likely to be added to the list of essentiality and where the best place to start is. To complicate matters, plant response to some elements is species related; not all plants respond equally to a particular element (Pallas and Jones 1978).

It should be remembered that since the beginning of time, plants have been growing in soils that contain all known elements. Those elements found in the soil and in the soil solution in a soluble form or forms as an ion can be taken into the plant by root absorption. This means that plants will contain most if not all those elements found in soil. Markert (1994) defined what he called the "reference plant composition" of plants, which included 26 elements that are not essential but are found in plants at easily detectable concentrations (Table 3.7). Some of these elements would be classed as "trace elements," since they are found in the plant's dry matter at low concentrations. This designation, however, can lead to some confusion, since the term "trace elements" was once used to identify what are defined today as the essential micronutrients. Some of these elements exist at fairly high concentrations in the plant depending on the level of their availability in soil.

Kabata-Pendias (2000) has given the approximate concentration of 18 nonessential elements found in plant leaf tissue, giving their range of sufficiency to toxicity to excess (Table 3.8). The question is which of the elements listed in Tables 3.6 and 3.7, irrespective of their found concentration in plants, would contribute, positively or negatively, to plant growth. A colleague and I (Pallas and Jones 1978) found that platinum (Pt) at very low levels (0.057 mg/L [ppm]) in a hydroponic nutrient solution stimulated plant growth for some plant species, but higher levels (0.57 mg/L ([ppm]) reduced growth for all species. The growth effects at the low level of Pt in solution varied considerably among nine plant species

Table 3.7 Trace Element Content of Markert's Reference Plant[a]

Trace element	mg/kg	Trace element	mg/kg
Antimony (Sb)	0.1	Iodine (I)	3.0
Arsenic (As)	0.1	Lead (Pb)	1.0
Barium (Ba)	40	Mercury (Hg)	0.1
Beryllium (Be)	0.001	Nickel (Hi)	1.5
Bismuth (Bi)	0.01	Selenium (Se)	0.02
Bromine (Br)	4.0	Silver (Ag)	0.2
Cadmium (Cd)	0.05	Strontium (Sr)	50
Cerium (Ce)	0.5	Thallium (Ti)	0.05
Cesium (Cs)	0.2	Tin (Sn)	0.2
Chromium (Cr)	1.5	Titanium (Ti)	5.0
Fluorine (F)	2.0	Tungsten (W)	0.2
Gallium (Ga)	0.1	Uranium (U)	0.01
Gold (Au)	0.001	Vanadium (V)	0.5

Source: Markert, B., 1994, in *Biochemistry of Trace Elements*, ed. D. C. Adriano, Z. S. Chen, and S. S. Yang, Science and Technology Letters, Northwood, NY.

[a] No data from typical accumulator and/or rejector plants.

(no response: radish and turnip; positive response: snap bean, cauliflower, corn, peas, and tomato; negative response: broccoli and pepper). It is the "stimulatory effect" of an element that needs to be investigated for those elements available in soils and soilless media that could be added to a hydroponic nutrient solution in order to benefit plant growth.

It was recognized by the early researchers that a "complete" nutrient solution should include not only the essential elements known at that time but also those that may be beneficial. Therefore, the A to Z micronutrient solution was developed (see Table 3.3). Those who may wish to explore the potential for discovery of additional elements that may prove essential for both animals and plants will find the books by Mertz (1981) and the articles by Asher (1991) and Pais (1992) interesting.

Four elements, Co, Ni, Si, and V, have been studied as to their potential essentiality for plants. Considerable research has been devoted to each of these elements, and some investigators feel that they are important (if not essential) elements for sustaining vigorous plant growth.

Cobalt (Co)

Cobalt is required indirectly by leguminous plants because this element is essential for the *Rhizobium* bacteria that live symbiotically in the roots, fixing atmospheric N_2 and providing the host plant much of its needed N. Without Co, the *Rhizobium* bacteria are inactive and the legume plant then requires an inorganic source of N as ions (as

Table 3.8 Approximate Concentrations of Micronutrients and Trace Elements in Mature Leaf Tissue

Micronutrient/trace element	Concentration (mg/kg dry weight)		
	Deficient	Sufficient or normal	Toxic or excessive
Antimony (Sb)	—	7 to 50	150
Arsenic (As)	—	1 to 1.7	5 to 20
Barium (Ba)	—	—	500
Beryllium (Be)	—	<1 to 7	10 to 50
Boron (B)	5 to 30	10 to 200	5 to 200
Cadmium (Cd)	—	0.05 to 0.2	5 to 30
Chromium (Cr)	—	0.1 to 0.5	5 to 30
Cobalt (Co)	—	0.02 to 1	15 to 50
Copper (Cu)	2 to 5	5 to 30	2 to 100
Fluorine (F)	—	5 to 30	50 to 500
Lead (Pb)	—	5 to 10	30 to 300
Lithium (Li)	—	3	5 to 50
Manganese (Mn)	15 to 25	20 to 300	300 to 500
Molybdenum (Mo)	0.1 to 0.3	0.211	0 to 50
Nickel (Ni)	—	0.1 to 5	10 to 100
Selenium (Se)	—	0.001 to 2	5 to 30
Silver (Ag)	—	0.5	5 to 10
Thallium (Tl)	—	—	20
Tin (Sn)	—	—	60
Titanium (Ti)	0.2 to 0.5	0.5 to 2.0	50 to 200
Vanadium (V)	—	0.2 to 1.5	5 to 10
Zinc (Zn)	10 to 20	27 to 150	100 to 400
Zirconium (Zr)	0.2 to 0.5	0.5 to 2.0	15

Source: Kabata-Pendias, H., 2000, *Trace Elements in Soils and Plants*, 3rd ed., CRC Press, Boca Raton, FL.

NO_3^- and/or NH_4^+) in the soil solution of a fertile soil. It is not clear whether the plant itself also requires Co to carry out specific biochemical processes. The irony of the relationship between *Rhizobium* bacteria and leguminous plants is that in the absence of sufficient inorganic N in the soil, which requires the plant to depend wholly on N_2 fixed by the *Rhizobium* bacteria, the plant will appear to be deficient in N, cease to grow, and eventually die if Co is not present. In the presence of adequate N in the rooting medium, colonies (nodules) of *Rhizobium* bacteria will not form on the plant roots.

Silicon (Si)

Plants that are soil grown can contain substantial quantities of Si, equal in concentration (percentage levels in the dry matter) to that of the major essential elements. Most of the Si absorbed (plants can readily absorb silicic acid, H_4SiO_4) is deposited in the plants as amorphous silica, $SiO_2 \cdot nH_2O$, known as opals. Epstein (1994) has identified six roles of Si in plants, both physiological and morphological. Reviewing 151 past nutrient solution formulations, Hewitt (1966) found that only a few included the element Si. Epstein (1994) recommends that Si as sodium silicate (Na_2SiO_3) be included in a Hoagland nutrient solution formulation at 0.25 mM. Morgan (2000) reported that hydroponic trials conducted in New Zealand resulted in yield improvements for lettuce and bean crops when the Si content in the nutrient solution was 140 ppm. Recent studies with greenhouse-grown tomato and cucumber have shown that, without adequate Si, plants are less vigorous and unusually susceptible to fungus disease attack (Belanger et al. 1995). Best growth is obtained when the nutrient solution contains 100 mg/L (ppm) of silicic acid (H_4SiO_4). The common reagent forms of Si added to a nutrient solution are either Na or K silicate, which are soluble compounds, while silicic acid is only partially soluble.

Silicon has been found to be required to maintain stalk strength in rice and other small grains (Takahashi, Ma, and Miyake 1990). In the absence of adequate Si in commercial production situations, these grain plants will not grow upright, with lodging resulting in significant grain loss. The problem of lodging has been observed primarily in paddy rice, where soil conditions may affect Si availability and uptake.

There can be confusion about this element, as the element silicon (Si) frequently is improperly referred to as silica, which is an insoluble compound, SiO_2.

Nickel (Ni)

Nickel is being considered an essential element for both legumes and small grains (i.e., barley), as Brown, Welsh, and Cary (1987) and Eskew et al. (1984) have shown that its deficiency meets the requirements for essentiality established by Arnon and Stout (1939) (see p. 34). Nickel is a component of the enzyme urease, and plants deficient in Ni have high accumulations of urea in their leaves. Nickel-deficient plants are slow growing and, for barley, viable grain is not produced. It is recommended that a nutrient solution contain a Ni concentration of at least 0.057 mg/L (ppm) in order to satisfy the plant requirement for this element, although its requirement for other than grain crops has not been established. Nickel is also related to seed viability; its deficiency in seed-bearing plants results in seeds that will not germinate.

Vanadium (V)

Vanadium seems to be capable of functioning as the element Mo in the N metabolism of plants, with no independent role clearly established for V. If Mo is at its sufficiency level (its requirement is extremely low—see Tables 3.1 and Table 3.2) in the plant, presence and availability of V are of no consequence.

Element substitution

There is considerable evidence that some nonessential elements can partially substitute for an essential element, such as Na for K, Rb for K, Sr for Ca, and V for Mo. These partial substitutions may be beneficial to plants in situations where an essential element is at a marginally sufficient concentration. For some plant species, this partial substitution may be highly beneficial to the plant. Despite considerable speculation, it is not known exactly how and why such substitutions take place, although similarity in elemental characteristics (atomic size and valance) may be the primary factor.

Visual plant symptoms of elemental deficiency or excess

In the literature, one can find descriptions of visual nutrient element deficiency and excess (toxic) symptoms as well as photographs showing visual symptoms at various stages of plant growth (Bennett, 1993). It should be remembered that visual symptoms may not appear similarly in all plants. In some instances, on-site visual symptoms may not be sufficiently distinct and therefore may be confusing to those unfamiliar with the techniques of diagnosis. In order to confirm a suspected insufficiency, it is desirable to have more than one individual observe the symptoms and to submit a properly collected plant (leaf) tissue sample for elemental laboratory analysis and interpretation (see Appendix C).

Visual symptoms as a result of elemental excess are not well identified for many of the essential plant nutrient elements. Some have said that symptoms of excess are not much different from those of deficiency for some elements, particularly for several of the micronutrients. Some elements can accumulate to levels far exceeding their physiological requirement but will not be detrimental to the plant. However, it is also known that when an element exists in the plant at a concentration far beyond its physiological requirement, such high levels may be "toxic" to the plant, interfering with specific or general physiological functions. Toxicity can also take place on the root surface if an element is at a particularly high concentration in the rooting medium or nutrient solution bathing the roots. An excess of one element may result in an imbalance among one or more other elements, resulting in a "toxic effect" in terms of root function and plant growth.

The combined influence of ions in solution will change the electrical conductivity (EC) of a solution surrounding the root, or a specific ionic balance may alter the pH of the surrounding solution. Therefore, the whole concept of excess can be confusing due to the varying factors associated with high concentrations of some elements in the rooting medium or the plant itself. Typical generalized symptoms of deficiency and excess are given in Table 3.9.

Table 3.9 Generalized Plant Nutrient Element Deficiency and Excess Symptoms

<table>
<tr><td colspan="2" align="center">Major elements</td></tr>
<tr><td colspan="2" align="center">**Nitrogen (N)**</td></tr>
<tr><td>Deficiency symptoms</td><td>Light green leaf and plant color; older leaves turn yellow and will eventually turn brown and die; plant growth is slow; plants will mature early and be stunted.</td></tr>
<tr><td>Excess symptoms</td><td>Plants will be dark green; new growth will be succulent; susceptible if subjected to disease, insect infestation, and drought stress; plants will easily lodge; blossom abortion and lack of fruit set will occur.</td></tr>
<tr><td colspan="2" align="center">**Ammonium (NH$_4$)**</td></tr>
<tr><td>Toxicity symptoms</td><td>Plants supplied with ammonium nitrogen (NH$_4$–N) may exhibit ammonium toxicity symptoms with carbohydrate depletion and reduced plant growth; lesions may appear on plant stems, along with downward cupping of leaves; decay of the conductive tissues at the bases of the stems and wilting under moisture stress; blossom-end fruit rot will occur and Mg deficiency symptoms may also appear.</td></tr>
<tr><td colspan="2" align="center">**Phosphorus (P)**</td></tr>
<tr><td>Deficiency symptoms</td><td>Plant growth will be slow and stunted; older leaves will have purple coloration, particularly on the undersides.</td></tr>
<tr><td>Excess symptoms</td><td>Excess symptoms will be visual signs of Zn, Fe, or Mn deficiency; high plant P content may interfere with normal Ca nutrition and typical Ca deficiency symptoms may appear.</td></tr>
<tr><td colspan="2" align="center">**Potassium (K)**</td></tr>
<tr><td>Deficiency symptoms</td><td>Edges of older leaves will appear burned, a symptom known as scorch; plants will easily lodge and be sensitive to disease infestation; fruit and seed production will be impaired and of poor quality.</td></tr>
<tr><td>Excess symptoms</td><td>Plant leaves will exhibit typical Mg and possibly Ca deficiency symptoms due to cation imbalance.</td></tr>
</table>

(Continued)

Table 3.9 Generalized Plant Nutrient Element Deficiency and Excess Symptoms
(Continued)

Calcium (Ca)

Deficiency symptoms	Growing tips of roots and leaves will turn brown and die; the edges of leaves will look ragged for the edges of emerging leaves will stick together; fruit quality will be affected and blossom-end rot will appear on fruits.
Excess symptoms	Plant leaves may exhibit typical Mg deficiency symptoms; in cases of great excess, K deficiency may also occur.

Magnesium (Mg)

Deficiency symptoms	Older leaves will be yellow, with interveinal chlorosis (yellowing between veins) symptoms; growth will be slow and some plants may be easily infested by disease.
Excess symptoms	Result is a cation imbalance with possible Ca or K deficiency symptoms appearing.

Sulfur (S)

Deficiency symptoms	Overall light green color of the entire plant; older leaves turn light green to yellow as the deficiency intensifies.
Excess symptoms	Premature senescence of leaves may occur.

Micronutrients

Boron (B)

Deficiency symptoms	Abnormal development of growing points (meristematic tissue); apical growing points eventually become stunted and die; flowers and fruits will abort; for some grain and fruit crops, yield and quality are significantly reduced; plant stems may be brittle and easily break.
Excess symptoms	Leaf tips and margins turn brown and die.

Chlorine (Cl)

Deficiency symptoms	Younger leaves will be chlorotic and plants will easily wilt.
Excess symptoms	Premature yellowing of the lower leaves with burning of leaf margins and tips; leaf abscission will occur and plants will easily wilt.

(Continued)

Table 3.9 Generalized Plant Nutrient Element Deficiency and Excess Symptoms
(Continued)

Copper (Cu)

Deficiency symptoms	Plant growth will be slow; plants will be stunted; young leaves will be distorted and growing points will die.
Excess symptoms	Iron deficiency may be·induced with very slow growth; roots may be stunted.

Iron (Fe)

Deficiency symptoms	Interveinal chlorosis on emerging and young leaves with eventual bleaching of the new growth; when severe, the entire plant may turn light green.
Excess symptoms	Bronzing of leaves with tiny brown spots, a typical symptom on some crops.

Manganese (Mn)

Deficiency symptoms	Interveinal chlorosis of young leaves while the leaves and plants remain generally green; when severe, the plants will be stunted.
Excess symptoms	Older leaves will show brown spots surrounded by chlorotic zones and circles.

Molybdenum (Mo)

Deficiency symptoms	Symptoms are similar to those of N deficiency; older and middle leaves become chlorotic first and, in some instances, leaf margins are rolled and growth and flower formation are restricted.
Excess symptoms	Not known and probably not of common occurrence.

Zinc (Zn)

Deficiency symptoms	Upper leaves will show interveinal chlorosis with whitening of affected leaves; leaves may be small and distorted, forming rosettes.
Excess symptoms	Iron deficiency symptoms will develop.

chapter four

The nutrient solution

Introduction

Probably no aspect of hydroponic growing is as misunderstood as the formulation and use of nutrient solutions. Most texts simply provide the reader with a list of nutrient solution formulas, preferred reagent sources, and the necessary weights and measures to prepare an aliquot of solution. Although such information is essential to prepare a nutrient solution properly, a soundly based understanding of its management is as important, if not more so, for successful growing. The complex interrelationships between composition and use are not understood by many formulators and most growers, and it is this aspect of nutrient solution management for which much of the literature unfortunately provides little or no help. In an article about a new growing machine for lettuce and herb production, called the "omega garden machine," the developers of the machine stated that "the hardest part is getting the plant food right and knowing how much to feed" (Simon 2004). This same thought can be echoed by many who have struggled with the selection and use of those nutrient solution formulations found in the hydroponic literature (Erickson 1990).

Poor yields, scraggly plants, high water and reagent costs—indeed, most of the hallmarks of a less than fully successful growing operation—can be directly linked to faulty formulations combined with the mismanagement of the nutrient solution (Gerber 1985; Jacoby 1995). There are, unfortunately, no absolute pat prescriptions or recipes that can be given to growers by any hydroponic advisor. Growers will have to experiment with their own systems, observing, testing, and adjusting until the proper balance between composition and use is achieved for their particular situation and specific plant species. However, what is surprising is that, for many instances, plants seem to be able to adjust, growing reasonably well, but not at their genetic potential. Genetic potential plant production requires precise management of the nutrient element environment of the rooting medium.

Although much is not known about how best to formulate and manage a nutrient solution, there are many good clues as to what should or should not be done. This chapter is devoted to an explanation of these clues. Growers using these clues will have to develop a scheme of management that best fits their environmental hydroponic system and plant growing conditions, experimenting with various techniques to obtain maximum utilization of the nutrient solution while achieving high crop yields of top quality.

The use of a particular nutrient solution formulation is based on three factors:

1. Hydroponic growing technique
2. Frequency and rate of nutrient solution dosing of plant roots
3. Plant nutrient element requirements

Water quality

All hydroponic growing systems require sizable quantities of relatively pure water. The best domestic water supplies or water for agricultural use frequently contain substances and elements that can affect (positively or negatively) plant growth. Even rainwater collected from the greenhouse covering may contain both inorganic and organic substances that can affect plant growth. In many parts of the United States, and indeed throughout the world, water quality can be a major factor for hydroponic use due to its content of various inorganic and organic substances. Therefore, a complete analysis of the water to be used for any type of hydroponic growing system is essential. The analysis should include inorganic as well as organic components if the water is being taken from a river, shallow well, or other surface sources. When taken from sources other than these, an inorganic elemental assay will be sufficient.

Natural water supplies can contain sizable concentrations of some of the essential elements required by plants, particularly Ca and Mg. In areas where water is being taken from limestone-based aquifers, it is not unusual for concentrations of Ca and Mg to be as high as 100 and 30 mg/L (ppm), respectively. Some natural waters will contain sizable concentrations of Na and anions, such as bicarbonate (HCO_3^-), carbonate (CO_2^{3-}), sulfate (SO_4^{2-}), and chloride (Cl^-). In some areas, B may be found in fairly high concentrations. Sulfide (S^-), primarily as iron sulfide, which gives a "rotten egg" smell to water, is found in some naturally occurring waters.

Suggested composition characteristics of waters suitable for use hydroponically as well as for irrigation have been published. Smith (1999) has given elemental maximums for water for hydroponic use (Table 4.1). Farnhand, Hasek, and Paul (1985) have established criteria for irrigation water based on salinity, electric conductivity (EC), total dissolved solids (TDS), and ion content (Table 4.2). Waters et al. (1972) have set the suitability of water for irrigating pot plants; their data are given in Table 4.3.

Surface or pond water may contain disease organisms or algae, which can pose problems. Algae grow extraordinarily well in most hydroponic culture systems, plugging pipes, and fouling valves. Filtering and/or other forms of pretreatment are required to ensure that the water used

Table 4.1 Common Compounds and Elements and Maximum Levels Allowable in Water for General Hydroponic Use

Element	Concentration, mg/L (ppm)
Boron (B)	<1
Calcium (Ca)	<200
Carbonates (CO$_3$)	<60
Chloride (Cl)	<70
Magnesium (Mg)	<60
Sodium (Na)	<180
Zinc (Zn)	<1

Source: Smith, R., 1999, *Growing Edge* 11(1):14–16.

to prepare the nutrient solution is free from these undesirable organisms and suspended matter.

Some form of water treatment may be necessary depending on what exists in the water supply. Simply filtering debris using either a sand bed or fine-pore filter may be at one end of the quality scale, while at the other extreme, sophisticated systems dedicated to ion removal by means of ion exchange or reverse osmosis may be found (Anon. 1997).

In hard-water areas, there may be sufficient Ca and Mg in the water to provide a portion or all of that required for the formulation. In addition, the micronutrient element concentration could be sufficient to preclude the need to add this group of elements to the nutrient solution. These determinations should be made only on the basis of an elemental analysis of the water. In

Table 4.2 Water Quality Guidelines for Irrigation

Characteristic	Degree of problem		
	None	Increasing	Severe
EC, dS/m[a]	<0.75	0.75 to 3.0	>3.0
TDS, mg/L[b]	<480	480 to 1920	>1920
Sodium (Na) sodium absorption ratio (SAR) value	<3	3 to 9	>9
Chloride (Cl) mg/L	<70	70 to 345	>345
Boron (B), mg/L	1.0	1.0 to 2.0	2.0 to 10.0
Ammonium (NH$_4$) and nitrate (NO$_3$), mg/L	<5	5 to 30	>30
Bicarbonate (HCO$_3$) mg/L	<40	40 to 520	>520

Source: Farnhand, D. S., Hasek, R. F., and Paul, J. L., 1985, Water quality, leaflet 2995. Division of Agriculture Science, University of California, Davis, CA.

[a] Electrical conductance.
[b] Total dissolved solids.

Table 4.3 Characteristics of High-Quality Irrigation Water

Characteristic	Desired level	Upper limit
Soluble salts (EC)	0.2 to 0.5 µS/cm	0.75 µS/cm for plugs; 1.5 µS/cm for general production
Soluble salts (total dissolved solids)	128 to 320 ppm	480 ppm for plugs; 960 ppm for general production
pH	5.4 to 6.8	7.0
Alkalinity (CaCO$_3$ equivalent)	40 to 65 ppm (0.8 to 1.3 meq/L)	150 ppm (3 meq/L)
Bicarbonates	40 to 65 ppm (0.70 to 1.1 meq/L)	122 ppm (2 meq/L)
Hardness (CaCO$_3$ equivalent)	< 100 ppm (2 meq/L)	150 ppm (3 meq/L)
Sodium (Na)	<50 ppm (2 meq/L)	69 ppm (3 meq/L)
Chloride (Cl)	<71 ppm (2 meq/L)	108 ppm (3 meq/L)
SAR[a]	<4	8
Nitrogen	<5 ppm (0.36 meq/L)	10 ppm (0.72 meq/L)
Nitrate (NO$_3$)	<5 ppm (0.08 meq/L)	10 ppm (0.16 meq/L)
Ammonium (NH$_4$)	<5 ppm (0.28 meq/L)	10 ppm (0.56 meq/L)
Phosphorus (P)	<1 ppm (0.3 meq/L)	5 ppm (1.5 meq/L)
Phosphate (H$_2$PO$_4$)	<1 ppm (0.01 meq/L)	5 ppm (0.05 meq/L)
Potassium (K)	<10 ppm (0.26 meq/L)	20 ppm (0.52 meq/L)
Calcium (Ca)	<60 ppm (3 meq/L)	120 ppm (6 meq/L)
Sulfate (SO$_4$)	<30 ppm (0.63 meq/L)	45 ppm (0.94 meq/L)
Magnesium (Mg)	<5 ppm (0.42 meq/L)	24 ppm (2 meq/L)
Manganese (Mn)	<1 ppm	2 ppm
Iron (Fe)	<1 ppm	5 ppm
Boron (B)	<0.3 ppm	0.5 ppm
Copper (Cu)	<0.1 ppm	0.2 ppm
Zinc (Zn)	<2 ppm	5 ppm
Aluminum (Al)	<2 ppm	5 ppm
Fluoride (F)	<1 ppm	1 ppm

Source: Whipker, B. E. et al., 2003, in *Ball Redbook: Crop Production*, vol. 2, 17th ed., ed. D. Hamrick, Ball Publishing, Batavia, IL.

[a] SAR = sodium absorption ratio = $Na^+/(Ca^{2+} + Mg^{2+})^{1/2}/2$.

addition, sufficient micronutrients supplied by the rooting medium (see pp. 92–97) may preclude their inclusion in a nutrient solution formulation.

Organic chemicals such as pesticides and herbicides, many of which are water soluble, can significantly affect plant growth if present even in low concentrations in a nutrient solution. Water from shallow wells or from surface water sources in intensively cropped agricultural areas should be tested for the presence of these types of chemicals. Treatment should be employed only if the chemical and/or physical composition of the water warrants. Obviously, financial and managerial planning must incorporate the costs of producing nutrient-pure water depending on the environmental conditions from which the water was taken. For example, it may be financially prudent to accept some crop loss from the use of impure water rather than attempting to recover the cost of water treatment. Treatment may be as simple and inexpensive a task as acidifying the water to remove bicarbonates (HCO_3) and carbonates (CO_3) or as expensive as complete ion removal by reverse osmosis.

Therefore, water samples should be submitted to an analytical laboratory for analysis before use, and the analysis should be repeated whenever a change in the water source is made. It is also advisable to have the initial nutrient solution assayed before its use to ensure that its composition is as intended.

Water pH

The pH of water can vary over a wide range; in addition, it can be difficult to determine accurately if the water contains few ions. For example, the pH of pure water is not an easily measurable determination, and if such water is exposed to air, its pH will vary depending on the amount of CO_2 adsorbed. The ratio of cations to anions, the types of ions, and their concentration in solution will determine a water's pH. For example, a saturated solution of $CaSO_4$ will be acidic because $CaSO_4$ is a salt of a strong acid and weak base. A solution of NaCl will be near neutral in pH because NaCl is a salt of a strong acid and strong base. Other comparisons can be made for other salts. Water with a mix of ions can have a wide pH range. In addition, the amount of dissolved CO_2 will play a role, less so in water with a high ion content versus water without a high ion content. Since most plants can grow well within a fairly wide acidic range in pH, pH adjustment may only be required when the water pH is at the extreme, or above neutrality (pH > 7.0).

Water and nutrient solution filtering and sterilization

Any suspended material in the water source should be removed by filtering through either a sand bed or a similar filter system (Anon. 1997). Suspended material may contain disease-carrying organisms, be a source for algae, or form precipitates with some elements in the used reagents when constituting the nutrient solution.

With continuous use in a closed recirculating system, the nutrient solution is altered with each passage through the root mass and/or rooting medium, not only by removal of elements by precipitation and plant root absorption but also through additions produced by the sloughing off of root material and substances contained in or incident to the rooting medium. As a result, the nutrient solution with each return to its storage tank will be physically and chemically changed due to the change in elemental content plus presence of suspended precipitates, microorganisms, and organic debris. In addition, there will be a considerable volume deduction.

For short-term use (less than 5 days), a change in physical or chemical composition of the nutrient solution may be of little consequence, although volume restitution is normally done. However, if the nutrient solution is to be used for an extended period of time (greater than 5 days), the replacement of spent elements must be made to extend its use, and filtering to remove suspended particles is also necessary.

Filtering the nutrient solution is not a common practice, nor is it recommended in most of the literature on hydroponics. The only exception would be water dispensed through a drip irrigation system, which must be free of suspended particles to prevent clogging of drippers.

The grower has a number of options to choose from for filtering the nutrient solution. Size, type, and installation requirements for a filtering system will vary depending on water volume, frequency of use, and quantity of material accumulating in the nutrient solution. Cartridge-type filters are recommended, as back-flushing is not generally possible or practical with most hydroponic systems, and cartridges can be easily replaced. Filtering devices should be placed in the outflow line leading to the growing bed from the supply reservoir or container. The coarser filter should be placed first in line, followed by the finer filter. Swimming pool-type filtering systems are capable of removing suspended particles of 50 μm and larger. Removal of particles below 50 μm requires the installation of a sophisticated filtering system, such as Millipore or similar type filters. Such a system is capable of removing substances that are microscopic in size (less than 1 μm). Thus, such a system removes not only large contaminants but also a number of disease organisms from the nutrient solution.

To provide some degree of control over microorganisms (bacteria, etc.), in addition to the use of a Millipore filter, the nutrient solution can be passed under ultraviolet (UV) radiation (Buyanovsky, Gale, and Degani 1981; Evans 1995). Ultraviolet sterilizers have proven to be effective in reducing microorganism counts when two 16 W lamps are placed in the path of the nutrient solution flowing at 13.5 L (3 gal) per minute, giving a total exposure of 573 J per square meter per hour. Another effective treatment is what is called "ozonation," bubbling ozone (O_3) through the nutrient solution—a treatment that will not alter its physical and chemical characteristics but will effectively sterilize the nutrient solution.

Weights and measures

Two sets of weights and measures are used in much of the hydroponic literature:

- English weight units—ounce (oz) and pound (lb)—and English volume measures—pint (pt), quart (qt), and gallon (gal)
- Metric weight units—gram (g) and kilogram (kg)—and metric volume measurements—cubic centimeter (cc), milliliter (mL), and liter (L)

British units are referred to as "non-SI" units and metric as "SI" units. A conversion table for converting non-SI to SI units and vice versa is found in Appendix A.

This text reports units as given in the source and provides converted units when appropriate. Although a considerable effort has been made to standardize units and measures worldwide, the hydroponic literature still uses a mix of units.

Nutrient solution formulations are generally based on making concentrates that are diluted and mixed together to give the nutrient solution that is applied to plant roots. The concentrates may be designated as part A, part B, etc., or as "macro" (containing the major elements) and "micro" (containing the micronutrients). In some instances, the concentrates may contain a mix of both major elements and micronutrients. The most common dilution rate from concentrate to final "to be used" nutrient solution is 1:100 (1 part concentrate to 100 parts water). However, other dilution rates may be used.

Nutrient solution reagents

What constitutes a nutrient solution is based on the reagents used in its formulation. The literature contains many nutrient solution formulations, but these can be confusing when the formulator uses a reagent name but does not give its elemental formula. For example, "potassium phosphate" is not sufficient, as there are two names for the same reagent: monopotassium phosphate or potassium dihydrogen phosphate (KH_2PO_4), and dipotassium phosphate or potassium monohydrogen phosphate (K_2HPO_4). The K and P contents of KH_2PO_4 are 30% and 32%, respectively, and for K_2HPO_4, 22% and 18%, respectively.

The other confusing factor is how many waters of hydration there are for the reagent specified. In general, most of the reagents used to formulate a nutrient solution have specific waters of hydration and may not pose a problem in identification since they are the usual commercial form, but this is not always the case for all reagents. For example, the usual commercial form of calcium nitrate, $Ca(NO_3)_2 \cdot 4H_2O$, has four waters of hydration, but $Ca(NO_3)_2$ is also available although it is not a commonly available or used form. The usual commercial form for Cu is copper sulfate, $CuSO_4 \cdot 5H_2O$, which has five waters of hydration, but copper sulfate, $CuSO_4$, without any waters of hydration is also available. The usual commercial form for Mn is $MnSO_4 \cdot 4H_2O$, which has four waters of hydration, although three other forms are available, with two, three, and five waters of hydration. The elemental composition of a reagent determines its formula weight and, in turn, will affect the weight of reagent used to make a nutrient solution. For example, the formula weight for $CuSO_4 \cdot 5H_2O$ is 249.71, while the formula weight for $CuSO_4$ is 159.63. Characteristics of commonly used reagents for formulating a nutrient solution are given in Tables 4.4 and 4.5.

The other issue is grade—whether fertilizer, pharmaceutical grade (US Pharmacopeia), or reagent; the differences among these grades mainly involve purity. Normally, fertilizer grade is sufficient for making a nutrient solution, although it is less pure than either USP or reagent grade forms, which are higher priced than fertilizer grade. One precaution is that the percentage of the element present in each grade may vary slightly and is usually lower in fertilizer grade materials, which also can contain low levels of related elements (for example, K fertilizers may contain Na).

Most hydroponic formulations are made using one, several, or all of the following chemical reagents:

- calcium nitrate [$Ca(NO_3)_2 \cdot 4H_2O$)]
- potassium nitrate (KNO_3)
- potassium dihydrogen phosphate (KH_2PO_4)
- magnesium sulfate ($MgSO_4 \cdot 7H_2O$)

With regard to the other essential elements—primarily the micronutrients—B is as either boric acid ($H_3BO_3 \cdot 5H_2O$) or borax ($Na_2B_4O_{24} \cdot 10H_2O$), and the sulfates for the elements Cu as copper sulfate ($CuSO_4 \cdot 5H_2O$), Fe as either ferrous sulfate ($FeSO_4$) or ferric ammonium sulfate [$FeSO_4(NH_4)_2SO_4 \cdot 6H_2O$], Mn as manganese sulfate ($MnSO_4 \cdot 4H_2O$), Mo as ammonium molybdate [$(NH_4)_6Mo_7O_{24} \cdot 4H_2O$], and Zn as zinc sulfate ($ZnSO_4 \cdot 7H_2O$)]. Normally, chlorine (Cl) is not specifically added to a nutrient solution formulation since this element is ever present in the environment and its requirement by most plants is very low.

Some of the micronutrients (mainly Fe) can be added as one of the chelates: FeEDTA or FeDTPA, for example. It has been found that EDTA (ethylenediaminetetra acetic acid) can be toxic to plants; therefore, DTPA (diethylene triamine pentaacetic acid) would be the desired chelate form

Table 4.4 Content of Plant Nutrients in Commonly Used Reagent-Grade Compounds

Reagent	Chemical formula	Content of nutrient in reagent-grade compound (%)
Ammonium nitrate	NH_4NO_3	N: 35.0
Ammonium sulfate	$(NH_4)_2SO_4$	N: 21.2; S: 24.)
Urea	$CO(NH_2)_2$	N: 46.6
Calcium nitrate	$Ca(NO_3)_2 \cdot 4H_2O$	N: 11.9; Ca: 17.0
Magnesium nitrate	$Mg(NO_3)_2 \cdot 6H_2O$	N: 10.9; Mg: 9.5
Potassium nitrate	KNO_3	N: 13.8; K: 38.7
Sodium nitrate	$NaNO_3$	N: 16.5
Monoammonium phosphate	$NH_4H_2PO_4$	N: 12.2; P: 27.0
Diammonium phosphate	$(NH_4)_2HPO_4$	N: 21.2; P: 23.S
Monocalcium phosphate	$Ca(H_2PO_4)_2 \cdot H_2O$	P: 24.6, Ca: l5.9
Dicalcium phosphate	$CaHPO_4$	P: 22.8; Ca: 29.5
Monopotassium phosphate	KH_2PO_4	P: 22.8; K: 28.7
Monosodium phosphate	$NaH_2PO_4 \cdot H_2O$	P: 22.5
Potassium chloride	KCl	K: 52.4
Potassium sulfate	K_2SO_4	K: 44.9; S: 18.4
Sodium sulfate	Na_2SO_4	S: 22.6
Calcium sulfate (gypsum)	$CaSO_4 \cdot 2H_2O$	Ca: 23.3; S: 18.6
Calcium carbonate	$CaCO_3$	Ca: 40.0
Magnesium carbonate	$MgCO_3$	Mg: 28 8
Magnesium sulfate (Epsom salts)	$MgSO_4 \cdot 7H_2O$	Mg: 9.9; S: 13.0
Ferrous sulfate	$FeSO_4 \cdot 7H_2O$	Fe: 20.1; S: 11.5
Manganese sulfate	$MnSO_4 \cdot H_2O$	Mn: 32.5; S: 19.0
Zinc sulfate	$ZnSO_4 \cdot 7H_2O$	Zn: 22.7; S: 11.2
Zinc oxide	ZnO	Zn: 80.3
Copper sulfate	$CuSO_4 \cdot 5H_2O$	Cu: 25.5; S: 12.8
Sodium borate (borax)	$Na_2B_4O_7 \cdot 10H_2O$	B: 11.3
Boric acid	H_3BO_3	B: 17.5
Sodium molybdate	Na_2MoO_4	Mo: 46.6

(Rengel 2002). In addition, the other heavy metal elements—Cu, Mn, and Zn—can also be added as their chelates. Chelates were developed primarily for use on alkaline soils or in organic soilless medium as a means of keeping the element in an available form in these two rooting environments. Therefore, they do not have a place in a nutrient solution formulation. Adding a chelated form of any element does not ensure its "availability" since, in a mixed elemental solution, the chelated bond can be easily broken and various combinations of chelated elements formed.

Table 4.5 Reagents, Formulas, Molecular Mass, Water Solubility, and Percent Element Composition of Commonly Used Reagents for Making Nutrient Solutions

Reagent	Formula	Solubility in cold water (15°C g/L)	Molecular mass	Percent of elements
Major elements				
Ammonium chloride	NH_4Cl	35	53.5	N 26
Ammonium nitrate	NH_4NO_3	1183	80.0	N 35
Ammonium sulfate	$(NH_4)2SO4$	706	132.1	N 21.2; S 24.3
Calcium chloride	$CaCl_2$	350	219.1	Ca 18.3
Calcium nitrate	$Ca(NO_3)_2 \cdot 4H_2O$	2660	236.1	Ca 17.0; N 11.9
Calcium sulfate	$CaSO_4$	2.41	172.2	Ca 23.3; S 18.6
Diammonium phosphate	$(NH4)_2HPO4$	575	132.0	N 21.2; P 23.5
Dipotassium phosphate	K_2HPO_4	1670	174.2	K 44.9; P 17.8
Magnesium nitrate	$Mg(NO_3)_2$	1250	256.4	Mg 9.5; N 10.9
Monoammonium phosphate	$NH_4H_2PO_4$	227	119.0	N 11.8; P 26
Phosphoric acid	H_3PO_4	5480	98	P 31
Potassium chloride	KCl	238	74.6	K 52.4
Potassium dihydrogen phosphate	KH_2PO_4	330	136.1	K 28.7; P 23.5
Potassium nitrate	KNO_3	133	101.1	K 38.7; N 13.8
Potassium sulfate	K_2SO_4	120	174.3	K 44.9; S 18.4
Sodium nitrate	$NaNO_3$	921	85.0	N 16.5
Urea	$CO(NH_2)_2$	1000	60.0	N 46.7
Micronutrients				
Ammonium molybdate	$(NH_4)_6Mo_7O_{24} \cdot 4H_2O$	430	1236	Mo 53
Boric acid	H_3BO_3	63.5	61.8	B 17.5
Copper sulfate	$CuSO_4 \cdot 5H_2O$	316	249.7	Cu 25.4
Iron (ferrous) sulfate	$FeSO_4$	156	278.0	Fe 20.1; S 11.5
Manganese chloride	$MnCl_2 \cdot 4H_2O$	1510	197.9	Mn 27.7
Manganese sulfate	$MnSO_4 \cdot 5H_2O$	1240	241.0	Mn 22.8
Manganese sulfate	$MnSO_4 \cdot H_2O$	985	169.0	Mn 32.4
Manganese sulfate	$MnSO_4 \cdot 4H_2O$	1053	223.0	Mn 24.6
Sodium borate (borax)	$Na_2B_4O_7 \cdot 10H_2O$	20.1	381.4	B 11.3
Sodium molybdate	$NaMoO_4$	443	205.9	Mo 46.6
Sodium molybdate	$NaMoO_4 \cdot 2H_2O$	562	241.9	Mo 39.6
Zinc sulfate	$ZnSO_4 \cdot 7H_2O$	965	287.5	Zn 22.7

Nutrient solution formulations

In this book, I have not made any changes in the format for the various nutrient solution formulas included in the text but have kept the format as presented by the formulator. Since no standard format exists, the volume of solution made, units (British or metric), and use instructions are as given by the formulation author. In some instances, the element content in a nutrient solution applied to a particular crop and/or specified hydroponic growing method is given, either with or without the formulation data. When formulation instructions are lacking, the user will have to make that determination.

While it is true that numerous formulations for preparing nutrient solutions have been published, their proper use relative to the growing system and needs for a specific plant species have been frequently overlooked. The basis for most hydroponic nutrient solution formulations comes from two formulas that appeared in the 1950 California Agricultural Experiment Station Circular 347 authored by Hoagland and Arnon (1950). This circular has been the most widely cited publication in all plant science literature. The scientific literature is full of hydroponic formulas that are identified as "modified Hoagland nutrient solutions" with little given that describes what was modified. What most readers do not know is that the Hoagland/Arnon nutrient solution formulations have use components—4 gal of nutrient solution per plant with replacement on a weekly basis. If any of these parameters is altered (i.e., volume of solution, number of plants, and frequency of replacement), plant performance can be significantly affected, a factor that is probably not fully understood or considered by those who recommend a particular nutrient solution formulation. The nutrient element contents for Hoagland/Arnon nutrient solution numbers 1 and 2 are given in Table 4.6.

Although a nutrient solution formula may be modified to suit particular requirements for its use, the critical requirements for proper management are either overlooked or not fully understood. The hydroponic literature is marked by comments on nutrient solution composition in terms of the concentration of the elements in solution, but is nearly devoid of instructions as to how the nutrient solution is to be used in simple management terms, such as the volume per plant and frequency of application. If the nutrient solution is recirculated, then the need for replenishment of specific elements prior to renewal should be specified.

When discussing questions regarding the use of a particular nutrient solution, Cooper (1988), developer of the Nutrient Film Technique (NFT) (Cooper 1979), remarked that "there is very little information available on this subject." In an interesting experiment, he obtained maximum tomato

Table 4.6 Hoagland and Arnon's Nutrient Solutions Number 1 and Number 2, Their Formulations, and Elemental Content

Stock solution	To use: mL/L
Solution no. 1	
1 M potassium dihydrogen phosphate (KH$_2$PO$_4$)	1.0
1 M potassium nitrate (KNO$_3$)	5.0
1 M calcium nitrate [Ca(NO$_3$)$_2$·4H$_2$O]	5.0
1 M magnesium sulfate (MgSO$_4$·7H$_2$O)	2.0
Solution no. 2	
1 M ammonium dihydrogen phosphate (NH$_4$H$_2$PO$_4$)	1.0
1 M potassium nitrate (KNO$_3$)	6.0
1 M calcium nitrate [Ca(NO$_3$)$_2$·4H$_2$O]	4.0
1 M magnesium sulfate (MgSO$_4$·7H$_2$O)	2.0
Micronutrient stock solution	
Boric acid (H$_3$BO$_3$)	2.86
Manganese chloride (MnCl$_2$·4H$_2$O)	1.81
Zinc sulfate (ZnSO$_4$·5H$_2$O)	0.22
Copper sulfate (CuSO$_4$·5H$_2$O)	0.08
Molybdate acid (H$_2$MoO$_4$·H$_2$O)	0.02
To use: 1 m/L nutrient solution	
Iron	
For solution no. 1: 0.5% iron ammonium citrate To use: 1 mL/L nutrient solution	
For solution no. 2: 0.5% iron chelate To use: 2 mL/L nutrient solution	

Element content of Hoagland/Arnon nutrient solutions (ppm)

Element	Hoagland no. 1	Hoagland no. 2
Nitrogen (NO$_3$)	242	220
Nitrogen (NH$_4$)	—	12.6
Phosphorus (P)	31	24
Potassium (K)	232	230
Calcium (Ca)	224	179
Magnesium (Mg)	49	49
Sulfur (S)	113	113
Boron (B)	0.45	0.45
Copper (Cu)	0.02	0.02
Manganese (Mn)	0.50	0.05
Molybdenum (Mo)	0.0106	0.0106
Zinc (Zn)	0.48	0.48

Source: Hoagland, D. R. and Arnon, D. I., 1950, The Water Culture Method for Growing Plants without Soil, circular 347, California Agricultural Experiment Station, University of California, Berkeley, CA.

plant growth when tomato plants were exposed to 60 L (13.3 gal) of nutrient solution per plant per week. Thinking that growth was enhanced by
the removal of root exudate due to the large volume of solution available
to the plants, he studied the relationship between root container size and
nutrient solution flow rate. He found that plant growth was affected prin-
cipally by the size of the rooting container and the volume of nutrient
solution flowing through the container, rather than by the removal of root
exudates. Cooper concluded that more fundamental research was needed
to determine the best volume of nutrient solution and flow characteristics
for maximum plant growth. He also observed that "the tolerance of nutrient supply was found to be very great."

   This observation seems to be in agreement with Steiner (1980), developer of the Steiner formula, who remarked that plants have the ability "to
select the ions in the mutual ratio favorable for their growth and development" if they are cultivated in an abundant nutrient solution flow.
Available evidence suggests that an advantage of flowing nutrient solution systems arises from the larger volume of nutrient solution available
to the plant, resulting in increased contact with the essential elements and
reduction in the concentration of inhibiting substances.

   Steiner (1961) has also suggested that only a handful of nutrient solution formulas are useful; at best, only one formulation would be sufficient
for most plants as long as the ion balance among the elements is maintained. Steiner felt that most plants will grow extremely well in one universal nutrient solution with the following percentage equivalent ratios of
anions and cations:

NO_3^-	50% to 70% of the anions
$H_2PO_4^-$	3% to 20% of the anions
SO_4^{2-}	25% to 40% of the anions
K^+	30% to 40% of the cations
Ca^{2+}	35% to 55% of the cations
Mg^{2+}	15% to 30% of the cations

   He also suggested that these ion concentration ratios may vary a bit
as follows:

NO_3^-	:	$H_2PO_4^-$	:	SO_4^{2-}
60	:	5	:	35
K^+	:	Ca^{2+}	:	Mg^{2+}
35	:	45	:	20

Steiner's (1980) thesis depends upon the assumption that plants can adjust to ratios of cations and anions that are not typical of their normal uptake characteristics, but that plants will expend much less energy if the ions of the essential elements are in proper balance as given previously. Steiner's thesis explains, in part, why many growers have successfully grown plants using Hoagland-type nutrient solution formulations, as plants are apparently able to adjust to the composition of the nutrient solution even when the ratios of ions are not within the range required for best plant growth. Steiner also suggested that the proper balance and utilization of ions in the nutrient solution are best achieved by using his universal nutrient solution formulas (Steiner 1984).

In contrast to the Steiner concept, Schon (1992) has mentioned the need to tailor the nutrient solution to meet the demands of the plant. Faulkner (1998a) gives instructions for the formulation of a modified Steiner solution as a complete nutrient solution, the composition of which is given in Table 4.7. Faulkner suggests that the Steiner nutrient solution is a "versatile 'complete' nutrient solution ideal for general hydroponic culture of a wide variety of greenhouse crops." A recipe for making 100 gal (378 L) of Steiner nutrient solution is given in Table 4.8.

The Hoagland and Arnon (1950) formulations provide another example of an imperfectly understood and improperly applied concept of nutrient element content for a nutrient solution formulation. The source of information for both of their nutrient solution formulas was obtained from the determination of the average elemental content of a tomato plant. They calculated the elemental concentration required based on one plant growing in 4 gal of nutrient solution, which was replaced weekly. Naturally, one might ask how these nutrient solution formulas would work if tomato is not the crop, the ratio of plant to volume of nutrient solution is greater or less than one plant to 4 gal, and the replenishment schedule is shorter or longer than 1 week. The simple answer is that we do not know, but experience suggests that their nutrient element formulations seem to "work" quite well in proving the essential elements required for normal plant growth to occur.

In the first comprehensive review of the hydroponic method, which covered more than a century, Hewitt (1966) gave the composition of over 100 nutrient solution formulas, giving their historical development beginning in 1860. In his book, Muckle (1993) lists 33 "general and historical formulas" covering the time period from 1933 to 1943 as well as formulas designed for use when growing specific plants, such as carnations, lettuce, strawberry, and tomato. Resh (1995) lists 36 formulas gathered from the literature covering the time period from 1865 to 1990; Jones (1998) published 22 major element formulas plus three micronutrient formulations gathered from the literature beginning in the late 1800s to more recent times. Yuste and Gostincar (1999) listed 34 unnamed formulas plus six

Table 4.7 Base Steiner Formula Consisting of Two 4 L Stock Solutions

Reagent	Formula	Grams (ounces)[a]
Stock solution, part 1		
Calcium nitrate (15.5% N, 19% Ca)	$Ca(NO_3)_2 \cdot 4H_2O$	364.9 (12.9)
Stock solution, part 2		
Monopotassium phosphate (22.7% P, 28.5% K)	KH_2PO_4	83.1 (2.9)
Potassium nitrate (13.75% N, 38% K)	KNO_3	55.0 (1.9)
Potassium sulfate (43% K, 17.5% S)	K_2SO_4	177.2 (6.3)
Iron EDTA (13% Fe)		8.7 (up to 14.5 g when feeding high Mn)
Zinc EDTA (14% Zn)		5.0 (0.17)
Copper EDTA (14.5% Cu)		0.3 (up to 1.3 g during bright sunny weather in spring to minimize fruit cracking problems)
Manganese EDTA (12% Mn)		3.1 (for tomatoes, increase up to 6.3 g during cloudy weather in winter to minimize deficiency problems)
Sodium molybdate (39.6% Mo)		0.1 (0.0035)
Borax or sodium borate (11.3% B)		3.3 (and up to 6.7 g)
Stock solution, part 3		
Epsom salt or magnesium sulfate (9.7% Mg, 13% S)	$MgSO_4 \cdot 7H_2O$	193.1 (6.8)

Elemental composition of a full-strength Steiner solution

Element	Concentration at 100% solution strength, mg/L(ppm)
Nitrogen (N)	170
Phosphorus (P)	50
Potassium (K)	320
Calcium (Ca)	183
Magnesium (Mg)	50

(Continued)

Table 4.7 Base Steiner Formula Consisting of Two 4 L Stock Solutions
(Continued)

Sulfur (S)	148
Iron (Fe)	3 to 4[b]
Manganese (Mn)	1 to 2[b]
Boron (B)	1 to 2
Zinc (Zn)	0.2
Copper (Cu)	0.1 to 0.5[c]
Molybdenum (Mo)	0.1

Source: Faulkner, S. P., 1998a, *The Growing Edge* 9(4):43–49.

[a] All are per 4 L of final volume in distilled water.
[b] Increase Mn to 2 ppm and iron to 4 ppm during cloudy weather.
[c] Increase Cu to 0.5 ppm during bright, sunny weather in spring to minimize fruit cracking.

Table 4.8 Recipe for 100 Gallons of Steiner Nutrient Solution

Reagent	Grams[a]	Ounces[a]	ppm of element
Potassium nitrate (KNO$_3$)	67	2.4	25 N, 65 K
Calcium nitrate [Ca(NO$_3$)$_2$·4H$_2$O]	360	12.7	147 N, 180 Ca
Potassium magnesium sulfate	167	5.9	80 K, 48 Mg, 37 S
Potassium sulfate (K$_2$SO$_4$)	140	5.0	154 K, 63 S
Chelated Fe (Fe 330 330–10% Fe)	11.5	0.4	3 Fe
Phosphoric acid (H$_3$PO$_4$) (75%)	50 mL		48 P
Micronutrient concentrate (see below)	200 mL	—	
Recipe for 16 L of micronutrient concentrate, which can be diluted to make 8000 gal			
Manganese sulfate (MnSO$_4$·4H$_2$O)	55.0		0.5 Mn
Boric acid (H$_3$BO$_3$)	86.5		0.5 B
Zinc sulfate (ZnSO$_4$·7H$_2$O)	16.8		0.2 Zn
Copper sulfate (CuSO$_4$·5H$_2$O)	24.2		0.2 Cu
Molybdenum trioxide (MoO$_3$) (66% Mo)	4.6		0.1 Mo

Source: Larsen, J. E., 1979, in *Proceedings of the First Annual Conference on Hydroponics: The Soilless Alternative*, Hydroponic Society of America, Brentwood, CA.

[a] Grams and ounces per 100 gallons of water.

named formulas (Hoagland; Turner and Herry; Ellis and Swaney; Mier-Schwart; Kiplin-Laurie; and Steiner) covering the time period from 1865 to 1960.

From these and other sources, it is interesting to note the ranges in elemental concentration in various nutrient solution formulas that are given in the books by Muckle (1993), Barry (1996), Jones (1997), and Yuste and Gostincar (1999), as well as in two articles published in *The Growing Edge* magazine (Table 4.9). Two possible explanations for why such ranges exist have to do with the method of hydroponic growing and the plant being grown. However, the ranges in element concentration seem unusually large and, perhaps, difficult to justify. The major and micronutrient concentration ranges and ionic forms in a typical nutrient solution are given in Table 4.10. Major element and micronutrient concentration ranges for a typical nutrient solution are given in Table 4.11.

General purpose/use formulations

Smith (1999) provides a nutrient formulation that he identifies as a "basic nutrient formula for general use" (Table 4.12). More recently, Morgan (2002a) has given the ingredients for what she has identified as a "general purpose hydroponic solution" (Table 4.12).

Plant species requirement adjustments of the nutrient solution

It is generally accepted that specific nutrient element plant species requirements exist that would be reflected in the elemental composition of a nutrient solution to be applied to that plant. Examples of what would be recommended for elements and plant species are given in Tables 4.14 and 4.15. In addition, nutrient solution composition adjustments may be made as the plant advances through its life cycle.

Nutrient solution control

In addition to the usual management considerations relating the costs for reagents and water, as well as the energy required to move the nutrient solution, the dispensing of a nutrient solution must be integrated into the operational plan of a hydroponic growing system. One of the major financial decisions involves balancing replenishment schedules against input costs and losses due to single-use dispensing systems and dumping versus multiple use with treatment requirements, and time and cost factors.

Table 4.9 Nutrient Element Concentration Range of Common Nutrient Solutions

Element	Range in concentration, ppm				
	Barry (1996)[a]	Jones (1997)[b]	Yuste/ Costincar (1999)[c]	10a (5)[d]	11a (5)[e]
Nitrogen (N)	70 to 250	100 to 200	47 to 284	140 to 300	100 to 200
			(NO$_3$–N) 14 to 33 (NH$_4$–N)		
Phosphorus (P)	15 to 80	30 to 50	4 to 448	31 to 80	15 to 90
Potassium (K)	150 to 400	100 to 200	65 to 993	160 to 300	80 to 350
Calcium (Ca)	70 to 200	100 to 200	50 to 500	100 to 400	122 to 220
Magnesium (Mg)	15 to 80	30 to 70	22 to 484	24 to 75	26 to 96
Sulfur (S)	20 to 200		32 to 640	32 to 400	
Boron(B)	0.1 to 0.6	0.2 to 0.4	0.1 to 1.0	0.06 to 1.0	0.4 to 1.5
Copper (Cu)	0.05 to 0.3	0.01 to 0.1	0.005 to 0.15	0.02 to 0.75	0.07 to 0.1
Iron (Fe)	0.8 to 6.0	2 to 12	Trace to 20	0.75 to 5.0	4 to 10
Manganese (Mn)	0.5 to 2.0	0.5 to 2.0	0.1 to 1.67	0.1 to 2.0	0.5 to 1.0
Molybdenum (Mo)	0.05 to 0.15	0.05 to 0.20	0.001 to 2.5	0.001 to 0.04	0.05 to 0.06
Zinc (Zn)	0.1 to 0.5	0.05 to 0.10	0.05 to 0.59	0.04 to 0.7	0.5 to 2.5

Sources: Barry, C., 1996, *Nutrients: The Handbook of Hydroponic Nutrient Solutions*, Casper Publications Pty Ltd., Narrabeen, NSW, Australia; Jones, J. B., Jr., 1997, *Hydroponics: A Practical Guide for the Soilless Grower*, St. Lucie Press, Boca Raton, FL; Yuste and Gostincar, eds., 1999, *Handbook of Agriculture*, Marcel Dekker, New York.

[a] Barry, C., 1996, *Nutrients: The Handbook of Hydroponic Nutrient Solutions*, Casper Publications Pty Ltd., Narrabeen, NSW, Australia.
[b] Jones, J. B., Jr., 1997, *Hydroponics: A Practical Guide for the Soilless Grower*, St. Lucie Press, Boca Raton, FL.
[c] Yuste, M. P. and Gostincar, J., eds., 1999, *Handbook of Agriculture*, Marcel Dekker, New York.
[d] Edwards, J., 1999, *The Growing Edge* 10(5):52–61
[e] Hankinson, J. 2000. *The Growing Edge*, 11(5):25

One set of terms used to describe two methods of nutrient solution management are "open" and "closed." An open system is one in which the nutrient solution is used only once in a one-way passage through the rooting vessel. In a closed system, after passing through the rooting medium or roots mass, the nutrient solution is collected and recirculated. These

Table 4.10 Basic Nutrient Formula for General Use

Reagent	Formula	Grams	Ounces
Bag A			
Calcium nitrate	$Ca(NO_3)_2 \cdot 4H_2O$	2000	70.60
Bag B			
Potassium nitrate	KNO_3	2275	80.25
Magnesium sulfate	$MgSO_4 \cdot 7H_2O$	1757	62.00
Potassium phosphate	KH_2PO_4[a]	878	31.00
Iron chelate (EDTA)		132	4.65
Manganese sulfate	$MnSO_4$	24.5	0.864
Boric acid	H_3BO_3	6.0	0.200
Copper sulfate	$CuSO_4 \cdot 5H_2O$	2.0	0.070
Zinc sulfate	$ZnSO_4 \cdot 5H_2O$	1.5	0.053
Ammonium molybdate	$(NH_4)_6Mo_7O_{24} \cdot 4H_2O$	0.35	0.0125

Source: Smith, R., 1999, *Growing Edge* 11(1):14–16.

To use: to 10 L (2.65 gal) of water, add 1 level teaspoon of bag A, stir until dissolved, and then add 1 level teaspoon of bag B and stir to dissolve.

[a] Assumed formula.

two means of nutrient solution management pose different requirements for managing a hydroponic growing system.

Faulkner (1998b) identified five nutrient solution characteristics given by Dr. John Larson (1979), emeritus professor of horticulture of Texas A&M University for "optimum production of disease-free greenhouse tomatoes, cucumbers, and other plant species":

- Using a properly balanced nutrient solution in the rooting zone
- Adjusting the pH to an optimum range favorable for plant use
- Having no ions present in toxic amounts or at levels that may interfere with other ions
- Holding the total salt concentration in the nutrient solution within 1500 to 4000 ppm
- Having a well-aerated rooting medium and maintaining the environmental temperature within the range of about 65°F to 75°F (18°C to 24°C)

All systems of nutrient solution management, whether open or closed, must lend themselves to precise control of the nutrient solution composition so that the concentration of elements can be varied in response to both known physiologic stages of development and the grower's sense of the condition of the plants.

Table 4.11 Optimum Range for the Essential Plant Nutrient Elements in a Nutrient Solution Formulation

Element	Optimum range[a] (ppm)
N (as nitrate)	80–150
N (as ammonium)[b]	10–30
P	15–30
K	100–150
Ca	150–200
Mg	50–80
S	50–100
B	0.1–0.6
Cu	0.01–0.10
Fe[c]	2–5
Mn	0.5–2.0
Mo[d]	0.05–0.15
Zn	0.1–0.5
Si[e]	>100

[a] These values are to be used as guidelines for making a general judgment of the suitability of a nutrient solution for use with most hydroponic growing systems and rooting media. The lower value is probably better than having the concentration at the high end of the range, particularly true when the use factors are high-volume applications and/or frequent applications.
[b] The presence of ammonium in a nutrient solution will enhance both N root absorption as well as utilization of N within the plant.
[c] Fe may not be needed in the formulation depending on the chemical composition of the rooting medium, as it may contain sufficient available Fe to meet the plant requirement.
[d] Mo may not have to be specifically added to a nutrient solution formulation since the plant requirement is very low and Mo may exist at a sufficient level in the rooting medium and/or exist as a companion element in the reagents used to formulate the nutrient solution.
[e] Silicon is not an essential plant nutrient element and there may be sufficient available Si in the rooting medium to meet the plant requirement.

When beginning, it is advisable to have the constituted nutrient solution assayed by an analytical laboratory to ensure that all of the elements in the nutrient solution are at the concentration specified in the formula.

It is very important in a closed recirculating hydroponic system to add water to the nutrient solution in order to maintain its original volume. In addition, some elements will have been removed along with the water, with these elements included in the makeup water. The question is how much of which element should be added. A common practice is to use an EC measurement of the nutrient solution as a means of determining what level of replenishment is needed. Surprisingly, this technique works fairly well. Unfortunately, such a measurement does not determine what differential change in elemental concentration may have taken place in order to

Table 4.12 General Purpose Hydroponic Nutrient Solution Formulation

Reagent	Formula	Quantity (g)
Part A		
Calcium nitrate	$Ca(NO_3)_2 \cdot 4H_2O$	13,110
Potassium nitrate	KNO_3	2557
Iron chelate		500
Part B		
Potassium nitrate	KNO_3	2557
Monopotassium phosphate	KH_2PO_4	3567
Magnesium sulfate	$MgSO_4 \cdot 7H_2O$	6625
Manganese sulfate	$MnSO_4 \cdot 4H_2O$	121
Zinc sulfate	$ZnSO_4 \cdot 7H_2O$	11
Boric acid	H_3BO_3	39
Copper sulfate	$CuSO_4 \cdot 5H_2O$	3
Ammonium molybdate	$(NH_4)_6Mo_7O_{24} \cdot 4H_2O$	1.02

Source: Morgan, L., 2002c, *The Growing Edge* 14(1):11.

To prepare: dissolve in two 26-gallon (100 L) stock solution tanks; to use: dilute 1:100 of both parts A and B, EC = 2.5, TDS = 1806.

Table 4.13 Recommended Major Element Concentrations in Nutrient Solutions by Crop

Crop	Major elements, mg/L (ppm)				
	Nitrogen	Phosphorus	Potassium	Calcium	Magnesium
Cucumber	230	40	315	175	42
Eggplant	175	30	235	150	28
Herbs	210	80	275	180	67
Lettuce	200	50	300	200	65
Melon	186	39	235	180	25
Pepper	175	39	235	150	28
Tomato	200	50	360	185	45

Source: Schon, M., 1992, in *Proceedings of the 13th Annual Conference on Hydroponics,* Hydroponic Society of America, ed. D. Schact, 1992, Hydroponic Society of America, ed. San Ramon, CA.

add back those element(s) that had been removed. Such a determination requires a complete elemental analysis of the nutrient solution.

The elements that are most likely to show the greatest change in the nutrient solution with use are N and K. One possible way would be to dilute the initial nutrient solution formula for the major elements only and add that as the makeup water, making this solution about one-quarter to one-third the strength of the original nutrient solution. Some experimenting

Table 4.14 Nutrient Solution Formulas for the Hydroponic Production of Tomato, Lettuce, and Rose

Reagent (fertilizer grade, g/100 L)	Tomato	Lettuce	Rose
Major elements (Schact, ed.)			
Calcium nitrate (15.5–0.0)	680	407	543
Magnesium sulfate	250	185	185
Potassium nitrate (13-0-44)	350	404	429
Potassium chloride (0-0-60)	170	—	—
Monopotassium phosphate (0-53-34)	200	136	204
Ammonium nitrate (33.5-0-0)	—	60	20
Micronutrients			
Iron chelate (10% Fe)	15.0	19.6	19.6
Manganese sulfate (28% Mn)	1.78	0.960	3.9
Boron (Solubor) (20.5% B)	2.43	0.970	1.1
Zinc sulfate (36% Zn)	0.280	0.552	0.448
Copper sulfate (25% Cu)	0.120	0.120	0.120
Sodium molybdate (39% Mo)	0.128	0.128	0.128

Source: van Zinderen Bakker, E. M., 1986, in *Proceedings 7th Annual Conference on Hydroponics: The Evolving Art, the Evolving Science*, Hydroponic Society of America, Concord, CA.

Table 4.15 Situations in Which Plants Increase or Decrease Nutrient Solution pH

pH change	Solution	Species
Increase	All N as NO_3^-	Most
Decrease	All N as NO_3^-	Some
Decrease	N as NH_4^+ or urea	Most
Decrease	All N as NO_3^-, no Fe	Fe efficient
Decrease	All N as NO_3^-, no P	Some
Decrease	No N	Most
Decrease	No N	N_2-fixing plants
Decrease	All N as NO_3^-	Shoot flushes, *Euonymus japonica*
Decrease	All N as NO_3^- solution depleted	Most?
Decrease	All N as NO_3^-	Shoots in dark, *E. japonica*

Source: Hershey, D. R., 1992, *Journal of Biological Education* 26(2):107–111.

and testing will be necessary to determine what that proper strength should be to avoid creating an ion imbalance by adding back too much or too little. The micronutrients should never be included in the makeup nutrient solution, thus minimizing the possible danger from excesses. Phosphorus is also an element that should possibly be excluded from the makeup solution.

Another factor that must be considered is what elements are being left behind in the rooting medium; the amount will vary depending on the rooting medium characteristics, the composition of the nutrient solution, and the frequency of recirculation. An important measurement and recommended procedure with some growing systems is periodically to take from an access port an aliquot of solution from the rooting medium, or of that flowing from it, and then determine its EC. At some designated EC reading, the rooting medium would then be leached with water to remove accumulated salts.

Anyone who has used gravel as a rooting medium, for example, may have noticed that with time a gray-white sludge (primarily precipitated calcium phosphate and calcium sulfate) begins to form, which may also entrap other elements, particularly the heavy metal (Cu, Fe, Mn, and Zn) micronutrients. Running one's hand through the gravel, it will become coated with a light gray colored sludge. The sludge can be a major source of elements for plant uptake irrespective of what is being added by means of the nutrient solution. This accumulation of sludge and its utilization by the plant can give rise to a gradual or sudden marked change in plant elemental content, which frequently is undesirable. Therefore, control of this type of accumulation needs to be part of the nutrient solution management program. I recommend that a sample of the growing medium be collected and analyzed as one would a soil (Jones 2001) and, based on the assay results, the nutrient solution formulation can be modified in order to minimize this accumulation. A grower, upon learning of a significant accumulation of some elements (mainly Ca, Mg, P, S, Fe, Mn, Cu, and Zn) in his gravel-sump growing system, altered his nutrient solution formulation, which consisted only of the elements K, N, and B in the nutrient solution being delivered to his tomato plants. This change resulted in significant savings in reagent costs and possibly avoided a potential nutrient element insufficiency.

pH of the nutrient solution

The "ideal pH" or "optimum pH range" for a nutrient solution stems mostly from data obtained from a combination of pH effects on nutrient element availabilities in soil or soilless organic media (Jones 2001). Argo and Fisher (2003) authored a comprehensive bulletin on "Understanding pH Management" that provides useful information on all aspects of pH measurement and pH effects on plants. Morgan (1998) gives the optimum pH range for 22 crops that can be hydroponically grown; the desired range in pH among these 22 species was between 5.0 and 7.5. In general, the range in pH suggested for most hydroponic solutions is between 5.8 and 6.5. Most nutrient solutions, when initially constituted, will have a pH between 5.0 and 6.0. There have been very few experiments conducted that would specifically define the "ideal pH" or "optimum pH range" for a nutrient solution, no matter how the solution is to be employed. It should be remembered that the pH of a nutrient solution is dependent on such factors as temperature, content of inorganic and organic ions and substances, types of ions present, and CO_2 content. Diurnal fluctuations in pH occur as the result of the changing solubility of CO_2 in the nutrient solution; however, these changes are usually not of sufficient magnitude to warrant daily adjustment. At any one point in time, the pH of a nutrient solution will oscillate about a point that can vary by as much as a 0.5 pH unit. Those who would recommend continuously monitoring and altering the pH of a nutrient solution may find this recommendation both costly and of no real benefit to the growing plant.

If the nutrient solution needs pH adjustment, the adding of an acid or alkali, as the case requires, to lower or raise the pH, respectively, can be made. A common procedure is to monitor the pH of a nutrient solution continuously when it is dispensed and inject either acid or alkali as required into the flowing stream of nutrient solution. Solutions of either sodium or potassium hydroxide (NaOH and KOH, respectively) are suitable alkalis for raising the pH. Ammonium hydroxide (NH_4OH) can also be used; however, it is more difficult to handle safely, and the addition of the NH_4^+ ion to the nutrient solution may not be desirable. Nitric (HNO_3), sulfuric (H_2SO_4), and hydrochloric (HCl) acids can be used for lowering the pH. An advantage or disadvantage for the use of HNO_3 would be the addition of the NO_3^- anion. Phosphoric acid (H_3PO_4) can also be used, but its use would add P, which might not be desirable.

Therefore, those acids and alkalis that contain one or more of the essential elements are less desirable for use than those that do not contain such elements. Thus, NaOH is the preferred alkali and either H_2SO_4 or HCl is the preferred acid, even though they contain essential elements, because their addition to the nutrient solution will have minimal effects. Commercially available pH control solutions for use in nutrient solutions are usually made from these reagents.

As stated earlier, nutrient solution pH and changes that can occur are influenced by many factors, such as N source (NO_3^- versus NH_4^+), nutrient deficiency (e.g., P-deficient plants cause pH to decline), plant species, and plant growth stage. Ikeda and Osawa (1981) observed that 20 different vegetable species showed a similar N source preference for either NO_3^- or NH_4^+-N when the pH of the nutrient solution was varied from 5.0 to 7.0. A considerable degree of pH control can be obtained by simply selecting a specific ratio of NO_3^- to NH_4^+ ions when the nutrient solution is initially prepared. If the ratio of NO_3^- to NH_4^+ is greater than 9 to 1, the pH of the solution tends to increase with time, whereas at ratios of 8 to 1 or less, pH decreases with time, as illustrated in Figure 4.1. Hershey (1992) also studied the influence of the NO_3^- and NH_4^+ content of a nutrient solution on its

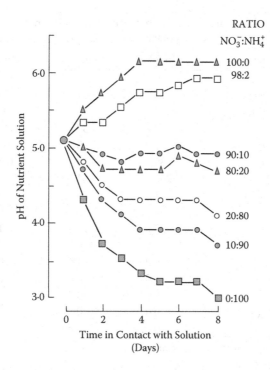

Figure 4.1 Effect of the ratio of nitrate to ammonium-nitrogen on the rate and direction of pH in nutrient solutions in contact with the roots of wheat (*Triticum aestivium*) plants. (Source: Trelease, S. E. and Trelease, H. M. 1935. *Science* 78:438–439.)

pH as affected by plant growth (Table 4.16). Hershey also observed "that NH_4^+ tends to be much more acidifying in solution than NO_3^- in alkalinizing; therefore a relatively small percentage of the N as NH_4^+ is effective in stabilizing the nutrient solution pH."

If the nutrient solution is constantly being adjusted upward to a neutral pH, it can interfere with the plant's natural ability to enhance its elemental ion-absorptive capability. Therefore, some have suggested that the pH of the nutrient solution should not be continuously adjusted but instead should be allowed to seek its own level naturally. This may be the desirable practice with those plant species sensitive to Fe when they are grown hydroponically as well as other plant species that have a particular elemental sensitivity that is pH related.

pH control of the nutrient solution may be akin to the nutrient solution filtering discussed earlier (see p. 54). It may be that more has been made about pH control and its potential effect on plants than can be justified from actual experience. Therefore, the requirement for pH control becomes a management decision, balancing benefits gained versus costs to control. It is obvious that extremes exist that the pH of the nutrient

Table 4.16 Composition of Some Common Nutrient Solution
Formulations[a]

Element	Hoagland	Nutri-Sol	Miracle-Gro
Nitrogen (total)	210	210	210
as nitrate (NO_3)	210	135	0
as ammonium (NH_4)	0	45	96
as urea	0	30	114
Phosphorus (P)	31	65	181
Potassium (K)	235	249	174
Calcium (Ca)	200	54	0
Magnesium (Mg)	48	9	0
Sulfur(S)	64	15	0
Iron (Fe)	5	2.3	1.4
Boron (B)	0.5	0.3	0
Manganese (Mn)	0.05	1.2	0.7
Zinc (Zn)	0.05	0.8	0.7
Copper (Cu)	0.02	0.8	0.7
Molybdenum (Mo)	0.01	0	0
Chlorine (Cl)	0.6	Trace	Trace

Source: Hershey, D. R., 1990, *The Science Teacher* 57:42–45.

[a] milligrams per liter.

solution should not be allowed to reach. What is needed to maintain the pH and prevent it from reaching those extremes may be academic, since those extremes are seldom reached with most nutrient solution formulas and their use.

Temperature of the nutrient solution

The temperature of the nutrient solution should never be less than the ambient air temperature, particularly in systems where plant roots are exposed to intermittent surges of a large volume of nutrient solution. On warm days, when the atmospheric demand on plants is high, root contact with nutrient solution below the ambient temperature can result in plant wilting, putting an undesirable stress on plants. Plant roots sitting in cool or cold nutrient solution cannot absorb sufficient water and elements to meet the demand of plant tops exposed to warm air and bright sunshine. Repeated exposure to cool nutrient solution results in slowed plant growth as well, as evidenced by poor fruit set and quality and delayed maturity. In such circumstances, it may be necessary to warm the nutrient solution to avoid this stress. On the other hand, warming the nutrient

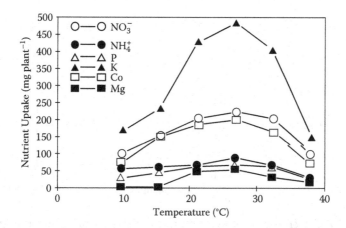

Figure 4.2 Influence of root temperature on major nutrient element uptake. (Source: Tindall, J. A. et al. 1990. *Journal of Plant Nutrition* 13:939–956.)

solution above the ambient temperature is not recommended and may do harm to plants.

Tindall, Mills, and Radcliffe (1990) found that for greenhouse hydroponically grown tomato, when the ambient air temperature was 70°F (21°C), maximum nutrient element uptake occurred for the major elements (Figure 4.2) and the micronutrients (Figure 4.3) when the nutrient solution temperature was 80°F (26.7°C). For maximum root and shoot growth, highest rate of shoot growth, and water uptake, the optimum root temperature was 77°F (25°C) (Figure 4.4).

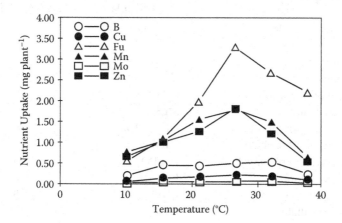

Figure 4.3 Influence of root temperature on micronutrient uptake. (Source: Tindall, J. A. et al. 1990. *Journal of Plant Nutrition* 13:939–956.)

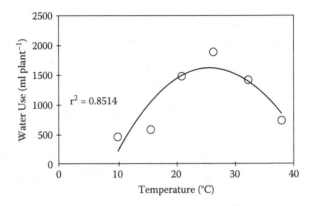

Figure 4.4 Influence of root temperature on tomato plant water use. (Source: Tindall, J. A. et al. 1990. *Journal of Plant Nutrition* 13:939–956.)

Electrical conductivity

The electrical conductivity (EC) of a nutrient solution as well as that retained in the rooting medium can significantly affect plant growth. Most nutrient solution formulas have a fairly low (<3.0 dS/m [mmhos/cm]) EC when initially made. For example, the Hoagland/Arnon number 1 nutrient solution given in Table 4.6 has an EC of 2.7 dS/m. The "salt effect" in a nutrient solution formulation can be minimized by selecting those compounds that have low salt indices (Table 4.17) when formulating the

Table 4.17 Relative Salt Index for Common Reagents Used for Preparing Nutrient Solutions

Reagent	Formula	Relative salt index
Ammonium nitrate	NH_4NO_3	104
Ammonium sulfate	$(NH_4)_2SO_4$	69
Calcium nitrate	$Ca(NO_3)_2 \cdot 4H_2O$	52
Calcium sulfate	$CaSO_4 \cdot 2H_2O$	8
Diammonium phosphate	$(NH_4)_2HPO_4$	29
Magnesium sulfate	$MgSO_4 \cdot 7H_2O$	44
Monoammonium phosphate	$NH_4H_2PO_4$	34
Monocalcium phosphate	$CaHPO_4$	15
Monopotassium phosphate	KH_2PO_4	30
Potassium chloride	KC	116
Potassium nitrate	KNO_3	73
Potassium sulfate	K_2SO_4	46
Sodium nitrate	$NaNO_3$	100
Urea	$CO(NH_2)_2$	75

nutrient solution. It is with use and/or reuse that a soluble salt problem arises. This problem develops when substantial quantities of water are removed at a very rapid rate from the nutrient solution when in contact with plant roots, as happens on warm, low-humidity days. This becomes particularly acute if the nutrient solution is being recirculated and the water loss due to evapotranspiration is not immediately replaced. If the water removed from the nutrient solution is not replaced, the EC of the nutrient solution will rise.

An EC measurement of the nutrient solution can also be used to determine the nutrient element replenishment level required to reconstitute the solution before reuse. From previous determinations, the amount of replenishment solution required to be added to the nutrient solution would be based on that EC measurement. Although this system of nutrient solution management has worked reasonably well, it does not take into account individual losses of elements from the nutrient solution by root absorption or retained in the rooting media. Therefore, replenishment based on an EC measurement may not fully reconstitute the nutrient solution in terms of its elemental composition.

In rockwool and perlite bag culture, for example, measurement of the EC of an aliquot of the retained solution in the rockwool slab or perlite or the effluent from them can be used to determine when leaching would be required to remove accumulated salts.

Oxygenation

The O_2 content of either a nutrient solution or the rooting medium will affect the rate of root activity and function, particularly the rate of water and nutrient element uptake. One of the major reasons that some NFT systems fail is due to the inability of the operating system to maintain sufficient O_2 in the ever expanding root mass in the NFT trough. This is also why the size and length of the NFT trough can be a critical factor, as at the end of the run little if any O_2 may remain in the passing nutrient solution. Attempts to oxygenate a nutrient solution prior to its introduction into the rooting medium have questionable value. Bubbling air or O_2 through a nutrient solution in an open environment will add some O_2 to the nutrient solution; the amount adsorbed depends on its temperature and ion composition. For example, even though a nutrient solution is saturated with O_2, its passage through a root mass or rooting medium can quickly strip all of the O_2 from it. Those plant roots in the initial contact position will benefit, but those further away will not.

The same NFT phenomenon will occur in vertical hydroponic culture growing systems (see p. 112), as the nutrient solution being delivered at the top of the rooting medium column will be stripped of its O_2 content as it moves down through the medium column.

Methods and timing of nutrient solution delivery

The nutrient solution can be already mixed at the desired application concentration for direct delivery to the rooting vessel, such as would be the case for ebb-and-flow, NFT, or aeroponics systems. The other method is to prepare elemental concentrates and, with the use of dosers (injectors; Figures 4.5 and 4.6), inject the appropriate aliquot of concentrate into a flowing water stream so that the final applied nutrient solution has the exact elemental composition specified in the nutrient solution formulation (Christian 2001). This is the method commonly used with drip irrigation systems.

Figure 4.5 Dosers (injectors) for dispensing stock solutions are shown in place over stock solution barrels. Dosers are adjustable so that they can deliver a specific aliquot of stock nutrient solution into a flow of water for constituting a nutrient solution for delivery to plants. One doser is used for each stock nutrient solution.

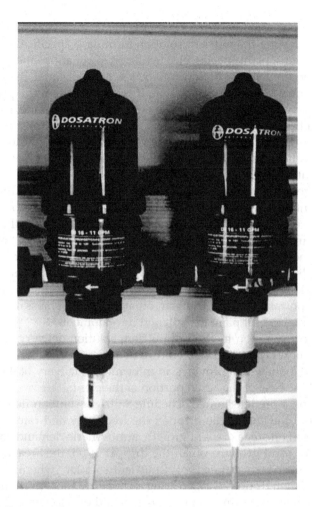

Figure 4.6 Dosatron dosers (injectors) for dispensing a stock nutrient solution.

The timing and techniques for delivering the nutrient solution to the rooting medium or plant roots will play a role in determining its composition. For a nutrient solution application schedule based primarily on the demand of the plants for water, the grower may be applying a nutrient solution when the nutrient element demand by the plants is already satisfied—that is, no additional nutrient elements at that specific time are needed. However, it is not the common practice simply to apply nutrient-free water to the plants, although such a capability would be desirable. As discussed earlier, with increasing frequency of application of a nutrient solution, the concentration of the nutrient elements in solution should be less. One could argue that on high atmospheric-demand days when plants are rapidly transpiring, both water and nutrient element requirements would be about equal in terms of what is being supplied by the nutrient solution (assuming the nutrient solution formula is specifically made for these conditions, plant species, etc.), while on lower atmospheric-demand days, the requirement for both would be less but still equal. By not adjusting the nutrient solution composition, it is assumed that the proper balance is achieved between water and nutrient element demand. Experience has shown that this concept is reasonably correct under most conditions. However, as the factors that relate to plant growth and development are better controlled, this assumed equal relationship between water demand and nutrient element need probably does not hold.

The size of the root mass is also a major factor that will affect water and nutrient element absorption (Barber and Bouldin 1984). As the root surface increases, the influx of water and nutrient elements through the roots also increases. In hydroponic systems, one might ask "how large must the root mass be to ensure that the demand for water and nutrient elements is met?" Unfortunately, no one has adequately made such a determination. There is some evidence which suggests that the root mass is not as important as root activity and that a large root mass may actually be detrimental to best plant growth and development.

The most common method of nutrient solution delivery used today with bag, bucket (pot), and slab culture hydroponic systems is by means of drip irrigation, which provides an intermittent delivery of the nutrient solution at the base of the aerial portion of the plant.

Based on a predetermined schedule, nutrient solution flows from its reservoir out the end of the dripper; the frequency and rate of flow are usually based on stage of plant growth, atmospheric-demand conditions, and stage of plant growth. When the dripper is on, the area around the point of delivery is saturated with nutrient solution; when it is off, the nutrient solution drains away, creating a changing root environment that may not be best for optimum plant growth and development. The draining of nutrient solution away from the point of introduction is considered desirable since air is drawn into the rooting medium, bringing with it O_2. Usually, sufficient nutrient solution is applied so that the area immediately under the dripper is leached, pushing any unused accumulated nutrient elements deeper into the bag, bucket (pot), or slab.

Normally, the bottom of the rooting vessel is open, allowing excess nutrient solution to flow out and the access holes or cuts are slightly above the bottom of the growing vessel so that a shallow depth of accumulated nutrient solution can be drawn on by the plant roots. Based on an analysis (usually a determination of EC) of a drawn solution sample from the medium or that being discharged, water will be periodically applied through the dripper to leach the growing medium of any accumulated salts of retained unused nutrient elements.

In an ebb-and-flow hydroponic system, the nutrient solution is pumped from a reservoir into the growing medium, flooding it with solution for a short period, and then the nutrient solution is allowed to flow out of the rooting medium back into the reservoir (see p. 108–110). This outflow of nutrient solution from the growing medium draws air into the rooting bed, providing a source of O_2. From the moist rooting medium, plants are able to obtain water and nutrient elements. Again, in such a system of nutrient solution delivery, the roots experience a changing environment, which may not be ideal for best plant growth and development, although plant performance is usually satisfactory with this hydroponic technique. In the rooting medium (the two common materials used for such systems are coarse sand or gravel), with time, an accumulation of unspent nutrient elements occurs in the form of precipitates, partially due to the drying of the rooting medium, which concentrates the elements once in solution. The precipitate is primarily a mixture of calcium phosphate and calcium sulfate, which will also occlude other elements in the applied nutrient solution (i.e., the micronutrients). Since the precipitates are not removed from the rooting medium by water leaching, the elements in the accumulating precipitate can become available for root absorption, thereby affecting plant growth and development.

For standing aerated systems (see p. 99), roots are suspended in a continuously aerated nutrient solution. Depending on the volume of nutrient solution versus number of plants, the nutrient solution elemental composition will be changing, therefore requiring periodic replenishing or replacement. A higher frequency of replenishment or replacement is needed when there is a large number of plants and/or a small volume of nutrient solution. Failure to replenish or replace when needed will result in poor plant performance.

In the NFT system, the nutrient solution flows down a channel occupied by plant roots (see pp.). As the distance from the point of introduction increases, the characteristics of the nutrient solution will significantly change: First the dissolved O_2 in the nutrient solution dissipates (Antkowiak 1993), followed by a change in the elemental composition of the solution. Therefore, the length of flow is critical. As the root mass increases, the nutrient solution will tend to flow over or around the root mass rather than through it, which will significantly affect plant performance.

In vertical growing columns (see p. 112), during the downward movement from the top of the column to the bottom, the nutrient solution will change considerably in elemental and O_2 content, as occurs in NFT systems. The length of the column and number of plants will determine the extent of change. Nutrient solution either applied on a timed schedule or based on atmospheric demand should be of sufficient volume to saturate the growing medium from top to bottom, creating an outflow at the bottom of the column.

For the aeroponic system, nutrient solution periodically bathes the roots with a fine mist of nutrient solution; the finer the mist is, the better the plant performance will be. Oxygen deficiency is not a problem, but the frequency of misting must be sufficient to keep the roots supplied with sufficient water to meet the transpiration demand of the plant. Under high atmospheric-demand conditions, a small reservoir of water or nutrient solution may be required at the base of the growing vessel so that the tips of the roots have access to this supply.

None of these commonly used nutrient solution delivery systems is without some undesirable aspect, although all are capable of delivering sufficient water and essential elements to sustain plant growth. The question is which system will work best in terms of efficient use of water and nutrient elements, resulting in high plant performance. The answer at this time is that none of them do, and the ideal delivery and utilization system has yet to be devised for commercial use.

Constancy

Maintaining the nutrient element status of the rooting medium at a constant level is not possible with the currently employed hydroponic growing systems. With each application of the nutrient solution to the rooting medium, the plant roots "see" a mix of nutrient elements from that remaining from previous nutrient solution applications and that being applied.

The benefits from maintaining a reasonable constancy of nutrient element concentration within the rooting medium were demonstrated in the following experiment. Snap beans were grown in pots in which perlite was the growing medium. The amount of water necessary to leach the entire perlite mass was determined as well as its retention volume. Each day prior to the hand application of an aliquot of nutrient solution, sufficient water was slowly applied to the perlite to replace what nutrient solution had been retained from the previous day's application. After allowing the water to drain from the perlite, an aliquot of nutrient solution was added based on what was needed to replace the previously applied water. This routine was followed every day during the experiment. Plant growth and pod yield were considerably greater than that obtained in previous experiments using the standing aerated nutrient solution method or in experiments where the nutrient solution was periodically dripped into the perlite-containing pot as needed in order to provide sufficient water as well as the required nutrient elements.

The only current hydroponic method that comes reasonably close to maintaining a constancy of water and nutrient elements is aeroponics (see p. 108). Unfortunately, aeroponics has not been widely adapted or used for a number of reasons. The work by the author on his GroSystem method comes close to maintaining consistency (see p. 114).

Programmable controllers

Numerous control systems are available for scheduling the dispensing of nutrient solutions. The controller may be a time clock on a preset timing schedule for dispensing a certain volume of nutrient solution or a system that is computer controlled, dispensing nutrient solution based on a demand determination, such as measured accumulated radiation. In addition, the controller may control other functions, such as adding a pH adjuster solution into the flow of nutrient solution or certain reagents to alter the composition of the nutrient solution. Since this technology is continuing to change as new devices are made available, it would not be appropriate to describe a system that could soon be obsolete.

The relationship that exists among the following three factors is not well understood:

- Elemental concentration of a nutrient solution
- Volume of nutrient solution applied with each irrigation
- Frequency of irrigations

Some authors have touched on these relationships. Cooper (1979) suggested modification in the concentration and delivery of a nutrient solution for use with the NFT method, the use of a so-called "drinking solution" (lower elemental concentrated solution), and sequencing between the application of a concentrated nutrient solution and use of water only or a lower concentrated nutrient solution application. Asher and Edwards (1978a, 1978b) found that rapidly moving large volumes of dilute nutrient solution formulations resulted in excellent plant growth, suggesting that plants exposed to an "infinite" volume of nutrient solution whose characteristics were not being altered by the growing plants were able to sustain plant growth. In fact, plant growth was more vigorous than that of plants grown in a currently recommended method of nutrient solution formulation and use.

Summary

There is no such thing as an "ideal" nutrient solution formulation; however, the Hoagland/Arnon formulations (see p. 60) will work with most plant species under a wide range of growing and environmental conditions. I recommend that, with this formulation, the Mg content be increased by 50%, the Zn content doubled, and the P content reduced 100% (Jones, 2012b).

The concept of balance among the cations and anions, as suggested by Steiner (1980, 1984), is worthy of further investigation. If a rapidly growing plant is placed into a standing aerated nutrient solution, that plant will quickly exhaust the nutrient solution of the K^+ and NO_3^- ions (primarily the monovalent cations and anions, and possibly B too), while most of the other elements in solution will change relatively little. This ease or lack of ease in uptake among the essential elements poses a challenge to the formulator to keep a nutrient solution in balance if its exposure to plant roots is lengthy. The ideal hydroponic growing system would be one in which the nutrient solution being supplied to the plant roots remains constant in its elemental composition within the rooting medium.

To some degree the NFT and aeroponic hydroponic growing methods approximate this condition of constancy. In a series of interesting experiments, Asher and Edwards (1978a, 1978b) observed that, if plants are grown in a rapidly moving nutrient solution of constant composition, the elemental concentration in the nutrient solution could be reduced significantly while plant growth remained normal. In fact, they found that most elements, particularly P, became toxic to plants unless reduced to concentrations (<2.6 mg/L [ppm]) considerably less than that recommended in most nutrient solution formulas. This indicates that plants grown in an infinite volume of nutrient solution so that plant uptake has no effect on the elemental concentration in solution would constitute what one could call the "ideal" hydroponic growing system.

It should be remembered that in medium-based hydroponic growing systems—and, possibly, to some extent in the NFT growing system—the plant is essentially drawing nutrient elements from three different nutrient element pools:

- Those currently supplied by applied nutrient solution
- Those remaining in the rooting media solution as ions (determined by an EC measurement)
- Those accumulating as precipitates

All of these pools can play major roles in determining the elemental content of the plant. This is probably one of the major factors contributing to nutrient element insufficiencies that will affect plant growth and fruit yield and quality. The objective of a nutrient element supply system should be to provide what is needed—no more and no less.

The quantity and balance approach developed by Geraldson (1963, 1982), although designed for soil-field-grown tomato, has potential application hydroponically. Such a system of approach has been found applicable to a soilless medium system for the production of a wide variety of greenhouse (Bruce et al. 1980) and garden (Jones 1980) vegetables. It is the basis for the *AquaNutrient* growing system developed by GroSystems (see www.GroSystems.com). This system approaches the "ideal," since plant roots are essentially exposed to a constant supply of both water and the essential elements.

Some of the issues that arise with the current formulations and use of nutrient solutions are as follows:

- Most nutrient solution formulations are not well balanced, particularly with regard to the major elements N and K.
- Total elemental concentration in most nutrient solutions is higher than can be justified in terms of meeting the plant requirement.
- Most nutrient element insufficiencies in plants are due to ion imbalances in the applied nutrient solution rather than to a deficiency of one or more elements.

The atmospheric demand should be a determinant of the total elemental concentration of a nutrient solution as well as a factor in determining the frequency of application (the higher the atmospheric demand is, the lower the element ion concentration should be in the nutrient solution with increased frequency of application).

There is justification for designing the nutrient solution delivery system so that only water can be applied, particularly during periods when the plant atmospheric demand is high. Also, being able to change the dilution ratio easily during the delivery of a nutrient solution would be a very useful factor.

The concentration of P in most nutrient solution formulations is about twice that needed and may be the primary cause for some plant nutrient insufficiencies among the micronutrients Cu, Fe, Mn, and, particularly, Zn.

The concentration of N in a nutrient solution may be the primary factor determining fruit yield and quality (the higher the N is, the lower the fruit yield and poorer the fruit quality will be). In general, the N content of a nutrient solution should be at the lower end of the recommended formulation amount and should be adjusted based on atmospheric demand—the higher the demand is, the lower the N concentration in the nutrient solution will be.

The ratio between K and Ca in a nutrient solution is probably a major factor determining fruit yield and quality. That nutrient solution elemental ratio for most plant species should be about 1 to 1. Equally important is the ratio of the three major cations—K^+, Ca^{2+}, and Mg^{2+}—with Mg the less competitive element. Magnesium deficiency is probably a commonly occurring deficiency, reducing plant growth without the occurrence of visual deficiency symptoms (see p. 43).

The use of chelated micronutrients may be the primary cause for deficiencies of the micronutrients Cu and, particularly, Zn in plants.

Insufficient Zn in most nutrient solution formulations may be the primary cause for low Zn levels in the plant. It is recommended that the Zn amount be double that specified in most nutrient solution formulations. It should be remembered that high P in a nutrient solution will inhibit Zn uptake as well as its distribution and function within the plant. The use of chelated Fe is also a contributor to lower Zn uptake and distribution within the plant.

The inclusion of NH_4–N in a nutrient solution formulation can enhance the uptake of NO_3–N, which can be either beneficial or detrimental. The amount of total N in a nutrient solution formulation can be reduced by 10% to 20% if 5% of the total N in the nutrient solution is NH_4.

The adjustment of the pH of a nutrient solution to a particular point is unjustified unless the pH is outside the desired range between 5.0 and 6.8. It should be remembered that the pH in the immediate area around plant roots is determined by the roots themselves.

The adjustment of a nutrient solution to a particular EC is probably not justified unless there is a compelling need to restrict water and nutrient element uptake.

The accumulation of elements as precipitates in the rooting medium, whether the medium is inorganic or organic, can have a significant effect on the plant's nutrition over time. Therefore, reducing the concentration of most elements—particularly, Ca, Mg, P, S, and Mn—is justified in the nutrient solution being applied over time.

The requirement for leaching a rooting medium due to the accumulation of unused elements can be significantly reduced by carefully adjusting the nutrient solution formulation and frequency of application as well as having the ability to apply only water for meeting high atmospheric-demand periods.

An EC measurement of nutrient solution flowing from the rooting medium or that exiting in the rooting medium is used to determine when the rooting medium requires water leaching. That requirement for leaching should be viewed as a warning signal that the quantity of nutrient elements being applied is greater than that required by the plants. This leaching requirement can be significantly reduced if greater care is used in formulating and applying (frequency and quantity) a nutrient solution; the ideal is that no water leaching is required. An elemental analysis of exiting or retained nutrient solution will indicate which elements are accumulating and provide guidance in reformulating the applied nutrient solution in order to minimize this accumulation.

The reuse of a rooting medium can pose a problem since that medium will start with a significant nutrient element charge from the accumulation of nutrient elements as precipitates that cannot be removed by water or even acid leaching.

In a closed nutrient solution system, the nutrient solution must be filtered and sterilized between applications.

An initially made nutrient solution should be assayed to determine its elemental content in order to ensure that all the elements are within the specifications of the formulation. Errors in selecting and weighing ingredients and mixing when preparing stock solutions can be easily made, and the malfunctioning of dosemonitors is not uncommon.

chapter five

Rooting media

Introduction

For rooting medium hydroponic culture systems, plants are rooted in an inorganic substrate with the nutrient solution applied by either periodically flooding the rooting media or by the use of a drip irrigation system. Some of the physical and chemical properties of commonly used inorganic substrates are given in Table 5.1. From this list, it can be seen that growers have a wide range of rooting media to choose from. In the past, course sand and pea gravel (materials that may have to be acid washed to remove unwanted substances and then, when once used, were usually discarded) were the media of choice. In addition, both substances have high volume weight and therefore require the use of sturdy rooting vessels. In addition, the flood-and-drain hydroponic growing system requires a rooting material that will not be moved within the rooting vessel when flooded with nutrient solution.

A slightly slanted sand table was at one time a commonly used hydroponic technique with the nutrient solution applied so that it flowed under the sand bed. However, more recently, perlite, rockwool, and coir have become the rooting media of choice with the nutrient solution being applied periodically by drip irrigation. All three substances have some similar physical properties as to their water-holding and aeration properties; are all, in general, inert; and have long-term physical and chemical properties. The elemental content and physiochemical properties are presented later in this chapter.

Perlite

Perlite is an amorphous volcanic glass that has a relatively high water-holding capacity, typically formed by the hydration of obsidian. It occurs naturally and has the unusual property of greatly expanding when heated sufficiently. It is an industrial mineral and a commercial product useful for its light weight after processing (www.wikipedia.com). Perlite has ample air space within the particles, thereby making it a desirable rooting material. It is inert and does not contain sufficient quantities of any of the essential plant nutrient elements. Perlite has been used in various ways: rooting plants in a bag of perlite, or the perlite is placed into pots or buckets of various forms and sizes. Normally, after use, the perlite is discarded.

Table 5.1 Characteristics of Inorganic Hydroponic Substrates

Substrate	Characteristics
Rockwool and stonewool	Clean, nontoxic (can cause skin irritation), sterile, lightweight when dry, reusable, high water-holding capacity (80%), good aeration (17% air-holding), no cation exchange or buffering capacity, provides ideal root environment for seed germination and long-term plant growth
Vermiculite	Porous, sponge-like, sterile material, lightweight, high water absorption capacity (five times its own weight), easily becomes waterlogged, relatively high cation exchange capacity
Perlite	Siliceous, sterile, sponge-like, very light, free-draining, no cation exchange or buffer capacity, good germination medium when mixed with vermiculite; dust can cause respiratory irritation
Pea gravel and metal chip	Particle size ranges from 5 to 15 mm in diameter; free draining; low water-holding capacity; high weight density, which may be an advantage or disadvantage; may require thorough water leaching and sterilization before use
Sand	Small rock grains of varying grain size (ideal size: 0.6 to 2.5 mm in diameter) and mineral composition; may be contaminated with clay and silt particles, which must be removed prior to hydroponic use; low water-holding capacity, high weight density; frequently added to an organic soilless mix to add weight and improve drainage
Expanded clay	Sterile, inert, range in pebble size of 1 to 18 mm, free draining, physical structure can allow for accumulation of water and nutrient elements, reusable if sterilized, commonly used in pot hydroponic systems
Pumice	Siliceous material of volcanic origin, inert, has higher water-holding capacity than sand, high air-filled porosity
Scoria	Porous, volcanic rock, fine grades used in germination mixes, lighter and tends to hold more water than sand
Polyurethane grow slabs	New material, which has a 75% to 80% air space and 15% water-holding capacity

Source: Morgan, L., 2003b, *Growing Edge* 15(2):54–66.

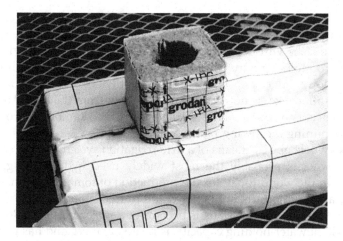

Figure 5.1 Rockwool slab wrapped in white polyethylene sheeting with a rock-wool cube placed on a cut opening in the sheeting.

Rockwool

Rockwool is a fibrous material produced from a mixture of volcanic rock, limestone, and coke; melted at 1500°C to 2000°C; extruded as fine fibers; and pressed into loosely woven sheets (Smith 1987). The sheets are made into slabs of varying widths (16 to 18 in. [15 to 46 cm]), normally 36 in. (91 cm) in length, and ranging in depth from 3 to 4 in. (5 to 10 cm). The slabs are normally wrapped with white polyethylene sheets as shown in Figures 5.1 and 5.2.

Figure 5.2 A typical rockwool slab with rockwool cubes spaced as would be appropriate for growing tomatoes.

The slabs are normally laid flat on a prepared floor surface, which is usually first covered by white polyethylene ground sheeting. Spacing among the slabs will depend on the configuration of the growing area and the crop to be grown. Once the slabs are set in place, cuts are made along the lower edge of each slab of the polyethylene slab covering on the bottom to allow excess nutrient solution to flow from the slab. An access hole is then cut on the top of the slab sheeting to accommodate a rockwool block containing a growing plant. Nutrient solution is then delivered to each rockwool cube by means of a drip irrigation system.

Rockwool is probably the most widely used hydroponic growing medium in the world today for the production of tomato, cucumber, and pepper, although efforts are being made to find an adequate substitute because disposal of used slabs is becoming a major problem. Rockwool has excellent water-holding capacity, is relatively inert, and has proven to be an excellent substrate for plant growth (Sonneveld 1989).

Coir

Coir is a natural fiber extracted from the husk of coconut—the fibrous material found between the hard, internal shell and the outer coat of a coconut (www.wikipedia.com). Coir is being recommended as a substitute for rockwool since it is an organic substance and can be more readily disposed of at the termination of its use as a rooting medium. Coir can be formed into blocks and slabs, such as those for rockwool, and used in much the same way.

Coir has much the same physical properties as rockwool, but it does contain both essential and nonessential elements— mainly sodium (Na). Therefore, coir may require water leaching to remove Na if it is high and thus could affect plant growth.

Elemental content of perlite, rockwool, and coir

These three rooting media are derived from naturally occurring substances and therefore will contain some, or many, of the essential plant nutrient elements. What portions of these elements are "available" for plant utilization is not generally known, nor easily determined due to their varying parameters of use. Experience, however, would suggest that a portion of the contained elements in a rooting medium can be available for root absorption. How that determination is made is the challenge.

Samples of perlite, rockwool, and coir were heated in *aqua regia* (mixture of concentrated hydrochloric and nitric acids) in order to bring them into solution. Perlite did not go into solution; therefore, no analytical data are given. However, rockwool and coir did and their obtained digests

Table 5.2 Elemental Content of Rockwool and Coir Brought into
Solution by *Aqua Regia* Digestion

Element	Rockwool (%)	Coir (%)
Phosphorus (P)	0.22	0.03
Potassium (K)	1.22	1.04
Calcium (Ca)	11.8	0.18
Magnesium (Mg)	3.0	0.23
Sodium (Na)[a]	0.81	
Sulfur (S)	0.20	0.59
Boron (B)	0.007	0.004
Copper (Cu)	0.01	0.001
Iron (Fe)	10.5	0.25
Manganese (Mn)	0.24	0.05
Zinc (Zn)	0.63	0.001

[a] Na content exceeded the analytical range of the spectrometer (probably greater than 5.00%).

were assayed for their elemental contents by ICP spectrometry. The assay results are given in Table 5.2.

Because they are natural products, these substances will have differences in elemental content depending on the composition of the source material; therefore, there is probably a "batch effect" that may or may not be significant. For rockwool, the chemical composition of the source mineral as well as the fluxing agent (limestone) will determine the final product's elemental content; for coir, the chemical environment associated with the production and processing of the coconut fiber will determine its element content.

Although these rooting media contain most of the essential plant nutrient elements, the question concerns what portion of these elements would be considered "available" for plant root absorption. Two extraction methods can be used for making that determination: water-equilibrium extraction or extraction using a soil extraction reagent (Jones 2001). The Mehlich No. 3 soil extraction method was chosen. The obtained extract was assayed for element content being expressed as pounds per acre (lb/A) so that the Mehlich No. 3 interpretation values could be applied. In addition to perlite, rockwool, and coir, two other commonly used rooting media—pinebark and peatmoss—were included; the results given in Table 5.3.

These soil extraction procedure results do not necessarily verify that that which is extracted defines what portion of the element content of the rooting medium is indeed "available" for root absorption, although it does provide a basis for comparison among rooting media. The assay results would suggest that all but perlite could be considered as a "fertile" soil. It

Table 5.3 Elemental Content of Rockwool, Perlite, and Coir Determined by
Mehlich No. 3 Extraction

Element	Rockwool (lb/A)	Perlite (lb/A)	Coir (lb/A)
Phosphorus (P)	54 (S)[a]	0.17 (D)	76 (S)
Potassium (K)	284 (S)	4.4 (D)	2340 (E)
Calcium (Ca)	2828 (S)	30.0 (D)	1,430 (S)
Magnesium (Mg)	774 (H)	7.6 (D)	866 (H)
Sulfur (S)	480 (H)	18.0 (D)	40 (S)
Boron (B)	2.0 (H)	0.90 (D)	4.0 (E)
Copper (Cu)	0.12 (S)	0.20 (D)	0.18 (S)
Iron (Fe)	2220.0 (E)	3.80 (S)	114.0 (E)
Manganese (Mn)	64.0 (E)	0.24 (S)	20.0 (H)
Zinc (Zn)	1.6 (S)	0.20 (S)	4.8 (H)

[a] D = deficient, L = low, S = sufficient, H = high, E = excessive.

may also suggest that using these rooting media, including the micronu-trients, would not be necessary in the selected fertilizer or nutrient solu-tion formulation. In addition, these results would suggest that one needs to match a nutrient solution formulation with the "available" elemental contents of the plant rooting media.

Next, what occurs when a nutrient solution is brought into contact with a rooting medium and allowed to come to equilibrium? To answer this question, a nutrient solution was added in an equal volume to rock-wool, perlite, and coir. The mixtures were stirred intermittently for 30 minutes, and then the liquid phase was removed by filtration and the fil-trate assayed for its elemental content by ICP spectrometry.

There are three possible outcomes:

1. No change in elemental content from that in the initial nutrient solution
2. An adsorption resulting in a decrease in concentration
3. A release resulting in an increase in the elemental content of the recovered nutrient solution

In Table 5.4, the elemental content of the nutrient solution (in parts per million, ppm) is given in the first column and the elemental content of the recovered nutrient solution following filtration after equilibrium—with rockwool, perlite, and coir, respectively—in the next three columns.

For perlite, the only elemental change was for the element Cu. For rock-wool, elemental change occurred for Ca—not surprising since Ca is a major constituent (see Table 5.3). Iron is also a major constituent, but no change occurred. The most significant changes occurred with coir, with increases in elemental contents for the elements P, K, Mg, Na, Fe, and B, and a decrease

Table 5.4 Elemental Interaction of Rockwool, Perlite, and Coir Brought into Equilibrium with a Nutrient Solution

Element	Nutrient solution (ppm)	Rockwool (ppm)	Perlite (ppm)	Coir (ppm)
Phosphorus (P)	82	80	88	140
Potassium (K)	276	272	288	678
Calcium (Ca)	198	282	192	50
Magnesium (Mg)	42	42	44	64
Sodium (Na)	15	15	19	218
Sulfur (S)	110	108	118	132
Boron (B)	0.26	0.27	0.29	0.70
Copper (Cu)	0.08	0.10	0.28	0.06
Iron (Fe)	1.1	0.8	0.7	2.7
Manganese (Mn)	0.72	0.78	0.76	0.60
Zinc (Zn)	0.28	0.28	0.28	0.27

in Ca. Some of this may be due to the fact that, in the interaction process, the filtrate was colored with colloidal organic material and that, if it had been removed, the elemental results might have been different.

These results suggest that rockwool and perlite could be considered inert in their interaction with an applied nutrient solution, while coir is not. The fact that both rockwool and coir are fibrous—therefore having a very large surface area—suggests that the likelihood of interaction would be high.

Based on the physiochemical nature of rooting media, the elements in an applied nutrient solution can potentially interact with the rooting media—being physically adsorbed or chemically bonded to form complexes and thereby resulting in their accumulation. To determine the degree of elemental accumulation that can occur, samples from a rockwool slab and perlite from BATO buckets were collected for elemental analysis following the growing of greenhouse tomatoes hydroponically using the drip irrigation system for nutrient solution delivery. Since, during the growing season, an accumulation of applied elements is observed as an increase in the electrical conductivity (EC) of the residue solution, growers are advised to monitor the EC of the retained solution and water leach when the retained solution reaches a certain EC level. For both collected samples, the growers were following a routine of periodic water leaching.

The gathered rockwool slab and perlite samples were first water leached and then extracted using the Mehlich No. 3 soil extractant; the results are expressed as pounds per acre (lb/A) so that the assay results can be interpreted using established Mehlich No. 3 interpretation values.

Table 5.5 Water-Soluble and Mehlich No. 3 Extraction Levels of Elements Found in Rockwool and Perlite after a Season of Use as Rooting Media in Greenhouse Production of Hydroponically Grown Tomatoes Using the Drip Irrigation Method

	Rockwool		Perlite	
Element	Water soluble (lb/A)	Mehlich No. 3 (lb/A)	Water soluble (lb/A)	Mehlich No. 3 (lb/A)
Phosphorus (P)	142	1066	32	384
Potassium (K)	3552	3382	439	513
Calcium (Ca)	231	5244	237	1071
Magnesium (Mg)	124	1177	43	65
Sulfur (S)	31	764	—	—
Boron (B)	0.5	1.7	0.29	0.6
Copper (Cu)	0.5	2.0	0.09	0.35
Iron (Fe)	—	3040+	—	—
Manganese (Mn)	2.0	104	0.34	8.0
Zinc (Zn)	1.6	26	0.08	2.0

Mehlich no. 3 extractable elements for unused rockwool and perlite are given in Table 5.4 and the assay results are given in Table 5.5.

The water-soluble assay results confirm that there is an accumulation of elements that remain in the rooting media in solution, while the Mehlich no. 3 extraction results indicate that another form of these elements exists, probably as precipitates of calcium sulfate and phosphate that either entrap other elements or form chemical complexes. The elements in these precipitates are probably "available" for root absorption since the plant root surfaces are acidic; when there is physical contact with a precipitate particle, some degree of dissolution may occur. These results also suggest that there needs to be an evaluation of the nutrient solution formulation so that both water-soluble and precipitate accumulation are minimized in order to avoid the potential for an occurrence of an essential element insufficiency from occurring in the growing crop.

Elemental content in a rooting medium can be a significant factor affecting the nutritional status of a growing crop. In addition, knowing what the elemental rooting media content is, one can match it with an appropriate fertilizer or nutrient solution formulation in order to avoid the potential of a plant nutrient insufficiency. An interaction can occur between an applied nutrient solution formulation and the rooting media, suggesting that matching media characteristics with a nutrient solution formulation are important in order to avoid an elemental insufficiency from occurring. Nutrient element accumulation in a rooting medium can be due to a residue increase as well as a possible formation of precipitates, combining to affect the nutritional status of the growing crop significantly. It also suggests that a nutrient solution formulation excessive in its elemental content will result in a significant accumulation of an element or elements, with the potential to affect the nutritional status of the growing crop adversely.

Being an "open" system, the nutrient solution is not recovered, and that delivered is sufficient for an excess flow from the cut openings on the bottom edge of the slab. Periodically, a solution sample is drawn from the slab and its EC determined; if it is found to exceed a certain level, the slab is leached with water. A pH measurement may also be made, and the nutrient solution composition may be changed if required. Normally, the elemental content of the slab-retained nutrient solution is not determined, although Ingratta, Blom, and Strave (1985) have given optimum and acceptable ranges for the solution of two crops (tomato and cucumber); the values are given in Table 5.6. These same values would also apply to other inert substrates.

Table 5.6 Optimum Concentrations and Acceptable Ranges of Nutrient Solution in a Rockwool Substrate[a]

	Tomato		Cucumber	
Determination	Acceptable	Optimum range	Acceptable	Optimum range
EC (µS/cm)	2.5	2.0 to 3.0	2.0	1.5 to 2.5
pH	5.5	5.0 to 6.0	5.5	5 to 6
Bicarbonate (HCO$_3$)	<60	0 to 60	60	0 to 60
Nitrate (NO$_3$)	560	370 to 930	620	440 to 800
Ammonium (NH$_4$)	<10	0 to 10	<10	1 to 10
Phosphorus (P)	30	15 to 45	30	15 to 45
Potassium (K)	200	160 to 270	175	140 to 270
Calcium (Ca)	200	160 to 280	200	140 to 280
Magnesium (Mg)	50	25 to 70	50	25 to 70
Sulfate (SO$_4$)	200	100 to 500	200	50 to 300
Boron (B)	0.4	0.2 to 0.8	0.4	0.2 to 0.8
Copper (Cu)	0.04	0.02 to 0.1	0.04	0.02 to 0.1
Iron (Fe)	0.8	0.4 to 1.1	0.7	0.4 to 1.1
Manganese (Mn)	0.4	0.2 to 0.8	0.4	0.2 to 0.8
Zinc (Zn)	0.3	0.2 to 0.7	0.3	0.2 to 0.7

Source: Ingratta, F. J., Blom, T. J., and Strave, W. A., 1985, in *Hydroponics Worldwide: State of the Art in Soilless Crop Production,* ed. A. J. Savage, International Center for Special Studies, Honolulu, HI.

[a] Milligrams per liter, parts per million, unless otherwise specified.

chapter six

Systems of hydroponic culture

Introduction

True hydroponics is the growing of plants in a nutrient solution without a rooting medium. Plant roots are either suspended in standing aerated nutrient solution or in a nutrient solution flowing through a root channel (known as Nutrient Film Technique [NFT]), or plant roots are sprayed periodically with a nutrient solution (known as aeroponics). This definition is quite different from the usually accepted concept of hydroponics, which has in the past included all forms of hydroponic growing. In the first section of this chapter, these three techniques of hydroponic growing will be discussed. In the second section, hydroponic systems using inorganic rooting media will be presented.

Another defining aspect of hydroponics is how the nutrient solution system functions—whether as an "open" system in which the nutrient solution is discarded after passing through the root mass or rooting medium, or as a "closed" system in which the nutrient solution, after passing through the root mass or rooting medium, is recovered for reuse.

Mediumless hydroponic systems

Standing aerated nutrient solution

This is the oldest hydroponic technique, dating back to those early researchers who, in the mid-1800s, used this method to determine which elements were essential for plants. Sachs in the 1840s and the other early investigators grew plants in aerated solutions and observed the effect on plant growth with the addition of various substances to the nutrient solution. This technique is still of use for various types of plant nutrition studies, although some researchers have turned to flowing and continuous replenishment nutrient solution procedures.

The requirements for the aerated standing nutrient solution technique are

- Suitable rooting vessel
- Nutrient solution

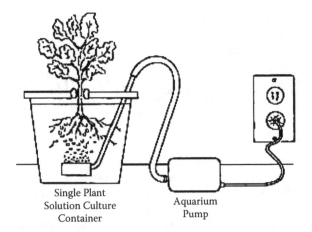

Single Plant
Solution Culture   Aquarium
Container   Pump

Figure 6.1 Standing aerated nutrient solution hydroponic growing system with an aquarium air pump attached to a line to a dispenser in the bottom of the growing vessel so that air can be bubbled into the nutrient solution.

- An air tube and pump in order to bubble air continuously into the nutrient solution, as shown in Figure 6.1

The bubbling air serves to add O_2 to the nutrient solution as well as stirring it. The commonly used formula is Hoagland/Arnon's (see Table 4.6 in Chapter 4, p. 60) or some modification of it as has been designed by Berry (1985), whose nutrient solution formula is given in Table 6.1, with the plant nutrient solution volume ratio of one plant per 2 to 4 gal (9 to 18 L) of nutrient solution.

The nutrient solution will require periodic replacement, usually every 5 to 10 days; the frequency is based on the number of plants and their size as well as the volume of nutrient solution. Water loss from the nutrient solution will need to be replaced daily, using either nutrient-free water (pure water) or a diluted (1/10 strength) nutrient solution, although there is the danger that any further additions of plant nutrient elements could alter the initial balance among the elements and adversely affect plants. It should also be remembered that with each day of use, the pH and composition of the initial nutrient solution will be altered by root activity and element uptake—changes that can have an adverse effect on plant growth. The question becomes "Should the pH and elemental content of the nutrient solution be restored daily to their original levels before replacement?" In most instances, adjustment other than water loss replacement is the practice normally followed.

Another aerated standing nutrient solution system has been described by Clark (1982); this technique has been used to study the elemental

Table 6.1 Stock Concentrates for Preparing a Nutrient Solution

| Reagent | Formula | Concentration | |
		g/L	Ounces/5 gallons
Stock concentrate 1			
Potassium nitrate	KNO₃	50.55	33.8
Potassium phosphate (mono)	Kh₂po₄	27.22	18.2
Magnesium sulfate	MgSO₄·7H₂O	49.30	32.9
Micronutrient concentrate		100 mL	64 fl oz
Micronutrient concentrate formulation			
Boric acid	H₃BO₃	2.850	1.90
Manganese sulfate	MnSO₄·H₂O	1.538	1.03
Zinc sulfate	ZnSO₄·7H₂O	0.219	0.15
Copper sulfate	CuSO₄·5H₂O	0.078	0.05
Molybdic acid	MoO₂·2H₂O	0.020	0.01
Stock concentrate 2			
Calcium nitrate[a]	Ca(NO₃)·4H₂O	118.0	78.8
Sequestrene 330 Fe[b]		5.0	3.3

Source: Berry, W.L., 1985, in *Proceedings of the 6th Annual Conference of Hydroponics,* Hydroponic Society of America, Concord, CA.

Notes: Approximate concentration of elements in final solution (mg/L, ppm): major elements: NO_3–N = 103, PO_4–P = 30, K = 140, Ca = 83, Mg = 24, SO_4–S = 32. Micronutrients: B = 0.25, Cu = 0.01, Fe = 2.5, Mn = 0.25, Mo = 0.005, Zn = 0.025. To use: 1:200 dilution in water

[a] Norsk hydro calcium nitrate is used, with the formula 5 Ca(NO_3) : 2 NH_4NO_3 : 10 H_2O; add only 88.8 g/L or 59 oz./5 gal.
[b] Mix the iron chelate thoroughly in a small amount of water before adding to the calcium nitrate.

requirements of corn and sorghum. Several plants are grown in half a gallon (2 L) of nutrient solution, with change schedules varying from 7 to 30 days depending on the stage of growth and plant species. The ratio of 8 to 1 of NO_3 to NH_4 in the nutrient solution is used to maintain some degree of constancy in pH. Clark's nutrient solution formula is given in Table 6.2. Although Clark's technique is primarily designed for corn and sorghum nutritional studies, his method of nutrient solution management could be successfully applied to other plant species.

The aerated standing nutrient solution method of hydroponic growing has limited commercial application, although lettuce and herbs have been successfully grown on Styrofoam sheets floating on an aerated nutrient solution (Figure 6.2). The plants are set in small holes in the Styrofoam, with their roots suspended in the nutrient solution. The sheets are lifted from the solution when the plants are ready to harvest.

Table 6.2 Composition of Nutrient Solution for Standing Aerated Growing System

Solution number	Reagent	Concentration (g/L)	Solution used (mL/L)	Cation	Anion
		Stock solution[a]		Full-strength nutrient solution (mg element/L)	
1[a]	Ca(NO$_3$)$_2$·4H$_2$O	270.0	6.6	Ca = 302.4	NO$_3$–N = 211.4
	NH$_4$NO$_3$	33.8		NH$_4$–N = 39.0	NO$_3$–N = 39.0
2	KCl	18.6	7.2	K = 70.2	Cl = 63.7
	K$_2$SO$_4$	44.6		K = 142.2	SO$_4$–S = 58.3
	KNO$_3$	24.6		K = 68.5	NO$_3$–N = 24.5
3	Mg(NO$_3$)$_2$·6H$_2$O	142.4	2.8	Mg = 37.8	NO$_3$–N = 43.6
4	KH$_2$PO$_4$	17.6	0.5	K = 2.5	P = 2.00
5[b]	Fe(NO$_3$)$_3$·9H$_2$O	13.31	1.5	Fe = 2.76	NO$_3$–N = 2.1
	HEDTA	8.68		Na = 4.48	HEDTA = 13.0
6	MnCl$_2$·H$_2$O	2.34	1.5	Mn = 0.974	Cl = 1.3
	H$_3$BO$_3$	2.04			B = 0.536
	ZnSO$_4$·7H$_2$O	0.88		Zn = 0.30	SO$_4$–S = 0.147
	CuSO$_4$·5H$_2$O	0.20		Cu = 0.076	SO$_4$–S = 0.038
	Na$_2$MoO$_4$·2H$_2$O	0.26		Na = 0.074	Mo = 0.155

Element	mg/L (ppm)	µM
	Final composition	
Calcium (Ca)	302	7540
Potassium (K)	283	7240
Magnesium (Mg)	37.8	1550
Nitrate-nitrogen (NO$_3$-N)	321	22,900
Ammonium-nitrogen (NH$_4$-N)	39.0	2780
Chlorine (Cl)	65.0	1940
Phosphorus (P)	2.00	65
Iron (Fe)	2.76	49
Manganese (Mn)	0.974	18
Boron (B)	0.536	50
Zinc (Zn)	0.300	4.6
Copper (Cu)	0.076	1.2
Molybdenum (Mo)	0.155	1.6
Sodium (Na)	4.56	200
HEDTA	13.0	47

(Continued)

Table 6.2 Composition of Nutrient Solution for Standing Aerated Growing
System (Continued)

Source: Clark, R. B., 1982, *Journal of Plant Nutrition*, 5(8):1003–1030.

[a] In each solution, the respective reagents were dissolved together in the same volume. Some of the reagents in solutions 1 to 4 may be combined to make fewer stock solutions if desired, but calcium reagents should be kept separate from sulfate (SO_4) and phosphate (PO_4) reagents. Combinations of the salts noted are for convenience.

[b] This solution was prepared by (a) dissolving the HEDTA [N.2(hydroxyethyl)ethylene-diamine-triacetic acid] in 200 mL distilled water + 80 mL 1 N NaOH; (b) adding solid $Fe(NO_3)_3 \cdot 9H_2O$ to the HEDTA solution and completely dissolving the iron salt; (c) adjusting the pH to 4.0 with small additions of 1 N NaOH in step (d) too rapidly to allow iron to precipitate. The HEDTA was obtained from Aldrich Chemical Co., Milwaukee, WI (catalog no. H2650-2).

Another reason why this system of growing hydroponically is not well suited for commercial application is that water and chemical use are quite high due to the requirement of frequent replacement. In addition, the composition of the nutrient solution is constantly changing, requiring monitoring and adjustment in order to maintain the pH and elemental ion balance and sufficiency concentration levels during the use period, which may range from 35 to 45 days, depending on the plant species grown and rate of plant growth. Temperature and root disease control are additional requirements if this method of growing is going to produce successful results.

Figure 6.2 Lettuce plants placed into openings in a Styrofoam sheet floated on a pool of nutrient solution.

Nutrient Film Technique

A significant development in hydroponics occurred in the 1970s with the introduction of the Nutrient Film Technique, frequently referred to by its acronym, NFT (Cooper 1976, 1979a, 1979b, 1988, 1995). Some have modified the name by using the word "flow" (Schippers 1979) in place of "film," as the plant roots indeed grow in a flow of nutrient solution. When Allen Cooper first publicly introduced his NFT system of hydroponic growing at the "Hydroponics Worldwide: State of the Art in Soilless Crop Production" conference (Savage 1985), it was heralded as the hydroponic method of the future (Edwards 1985). It was, indeed, the first major change in hydroponic growing techniques since the 1930s. Cooper and his colleagues discussed their experiences with this method, which left those in attendance with the belief that the science of hydroponics had made a major step forward.

Experience has shown, however, that the NFT method does not solve the common problems inherent in most hydroponic growing systems. However, this did not deter its rapid acceptance and use in many parts of the world, particularly in Western Europe and England. But its future continues to be highly questionable unless reliable means of disease and nutrient solution control are found. A change in the design of the trough has been suggested by Cooper (1985), from the "U" shape to a "W" (called a divided gully system), in which the plant base sits on the top of the "W" center with the roots divided down each side of the "W." A capillary mate is placed on the inverted "V" portion of the "W" to keep the roots moist with nutrient solution.

There are a number of advantages to this redesign of the NFT single-gully system as initially proposed by Cooper (1976, 1979a, 1979b). A portion of the plant roots—that on the inverted "V"—is in air; a portion of the roots lies on a moist surface (capillary matting), which provides for better oxygenation of the roots; and the remaining root mass is divided into two channels, which should minimize the problems associated with a large mass of roots in a single channel. It is now possible to use two different irrigation systems by flowing water or various types of nutrient solutions down either channel. Unfortunately, the NFT channel system has now been made more complicated in design, and it is uncertain whether this change will significantly improve plant performance. Cooper (1996) has published a revision of his 1976 book on NFT in which he recognizes some of the problems that can occur with this technique of hydroponic growing.

Simply put, in the NFT system, plant roots are suspended in a trough, channel, or gully (trough will be the word used from this point on) through which a nutrient solution passes. The trough containing the plant roots is set on a slope (usually about 1% to 2%) so that the nutrient solution introduced at the top of the trough can flow from the top to the lower end

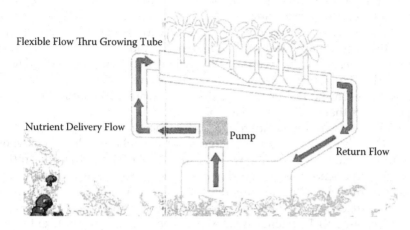

Flexible Flow Thru Growing Tube

Nutrient Delivery Flow

Pump

Return Flow

Figure 6.3 Typical arrangement for a closed NFT system in which a nutrient solution is pumped from a storage tank into the sloping NFT trough and then, by gravity, flows back into the storage tank.

by gravity at a recommended flow rate of one-quarter of a gallon (1 L) per minute (Figure 6.3). As the root mat increases in size, the volume rate down the trough diminishes. As the nutrient solution flows down the trough, plants at the upper end of the trough will reduce the O_2 and/or elemental content of the nutrient solution, a reduction that can be sufficient to affect growth and development of plants significantly at the lower end.

Furthermore, as the root mat thickens and becomes denser, the flowing nutrient solution tends to move over the top and down the outer edge of the root mat, reducing its contact within the root mass itself. This interruption in flow results in poor mixing of the current flowing nutrient solution with water and elements left behind in the root mat from previous nutrient solution applications. One of the means for minimizing these effects is to make the trough no longer than 30 ft (9 m) in length. In addition, the trough can also be made wider, which can be more accommodating for root growth with longer term crops.

One of the major advantages of NFT is the ease of establishment and the relatively low cost of construction materials. The design of NFT troughs and materials suitable for making troughs is discussed by Morgan (1999b) and Smith (2004). A trough can be simply formed by folding a wide strip of polyethylene film into a pipe- or triangular-like shape. The polyethylene film may be either white or black but must be opaque to keep light out. If light enters the trough, algae growth becomes a serious problem. The polyethylene sheet is pulled around the plant stem and closed with pins or clips, forming a lightproof, pipe-like rooting trough. If the trough is formed from strips of polyethylene film, it can be discarded after each crop, thus necessitating sterilization only of the permanent piping and nutrient solution storage tank.

Most troughs in use today are made of various plastic materials; the requirements are opacity, structural strength, and ultraviolet (UV) resistance. The design of the trough (width, height, and form) is usually determined by the crop to be grown. Lack of structural strength can lead to unevenness in the trough bottom that allows nutrient solution to lie in depressions that can lead to anaerobic conditions.

The plants are set in the trough at the spacing recommended for that crop. Usually, plants are started in germination cubes made of rockwool or similar material. The cube with its started plant is set directly in the trough. Experience has shown that the germination cube should not be made of materials that disintegrate with time. A durable germination cube helps keep the plant set in place in the NFT trough.

Normally, NFT systems are closed systems; that is, the nutrient solution exiting the end of the trough is recovered for reuse. Bugbee (1995) discusses the requirements for the management of recirculating hydroponic growing systems. The addition of makeup water, the need for reconstituting the pH and nutrient element content, filtering, and sterilization are procedures that need to be established. An open system would mean that the nutrient solution exiting the trough is discarded, which is costly in terms of water and reagent use as well as posing a problem for proper disposal (Johnson 2002c).

If the NFT system is operated as a closed system (i.e., the nutrient solution is recirculated a number of times before being discarded), Cooper (1979a) has recommended the use of a special nutrient solution—referred to as the "topping-up solution"—to be added to the starting solution to maintain its composition during use. Normally, the nutrient solution is monitored by periodic electrical conductivity (EC) measurements, which determine the appropriate times to add makeup (or topping-up) nutrient solution to maintain the initial volume and when to dump and make a new batch of nutrient solution.

The timing to flow the nutrient solution down the NFT trough varies. One practice is to flow the nutrient solution intermittently down the trough on an "on–off" cycle or by a "half-on, half-off" circulation period; a more sophisticated system is based on timing recirculation on the accumulation of incoming radiation. For example, when 0.3 mJ/m^2 of light energy has accumulated, the nutrient solution is flowed down the trough for 30 minutes; the time and length are also affected by the crop and its stage of growth. Such systems are coming into wider use because they have proven to be successful in producing better and higher yielding tomato and cucumber crops.

The NFT principle has also been applied to smaller growing units for home garden use. For example, one such application for vegetable

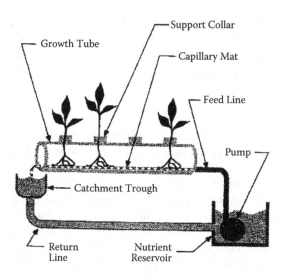

Figure 6.4 Typical arrangement for a hobby-type closed NFT system from which the nutrient solution is pumped from a reservoir into a sloping grow tube, with nutrient solution flowing down the tube by gravity and back into the nutrient solution reservoir. (Source: van Patten, 1992. *Growing Edge* 3(3): 24–33, 48–51.)

growing places sand-filled Styrofoam cups in access holes in PVC pipes. The nutrient solution circulates through the pipe on a timed schedule. This system has the unique feature of easy removal of plants by lifting the Styrofoam cup from its access hole. A typical arrangement for the hobby-type NFT system is illustrated in Figure 6.4.

Disease control can be difficult because a disease organism entering an NFT system will be quickly carried from one plant to another in the trough and from one trough to another if the nutrient solution is recirculated and not sterilized. Therefore, the same precautions are required as for any closed recirculating nutrient solution growing system. In warm climatic areas, the fungus *Pythium* is the major organism affecting plants grown in NFT systems. *Pythium* does not seem to be a serious problem when the temperature of the nutrient solution is maintained at less than 70°F (25°C).

Root death is another problem in NFT installations and may be the result of a lack of O_2 in the root mass (Antkowiak 1993). Recently, it has been suggested that concern is greater than justified, in as much as root death is a natural physiological phenomenon brought on by competition within the plant for carbohydrates. During periods of high demand for carbohydrates (primarily at fruiting or during times of stress), some roots will die, but when stress is relieved, plant tissue regains an adequate carbohydrate supply and new roots will appear. As long as most of the roots in the mat are alive, some suggest that little attention should be paid to root death. This phenomenon probably occurs in all systems of growing; it is clearly visible in NFT but not as easily seen when roots are growing in an inorganic or organic medium.

Aeroponics

A promising hydroponic technique for the future was thought to be aeroponics, which is the application of water and essential plant elements by means of an aerosol mist bathing the plant roots (Nichols 2002). One of the significant advantages of this technique compared to flowing the nutrient solution past the plant roots is aeration, as the roots are essentially growing in air. The technique was designed to achieve substantial economies in the use of both water and essential plant nutrient elements. The critical aspects of the technique are the character of the aerosol, frequency of root exposure, and composition of the nutrient solution. Several methods have employed a spray of the nutrient solution rather than a fine mist; droplet size and frequency of exposure of the roots to the nutrient solution are the critical factors.

Continuous exposure of the roots to a fine mist gives better results than intermittent spraying or misting. In most aeroponic systems, a small reservoir of water is allowed to remain in the bottom of the rooting vessel so that a portion of the roots has access to a continuous supply of water. The composition of the nutrient solution would be adjusted based on the time and frequency of exposure of the roots to the nutrient solution.

One of the applications of aeroponics is for herbs when the root is the portion of the plant harvested. A commercial application of aeroponics is the AeroGarden (www.areogarden.com).

Rooting medium hydroponic systems

In the culture systems described in this section, plants are grown in some type of inorganic rooting medium with the nutrient solution applied by flooding or drip irrigation. The physical and chemical properties of commonly used inorganic substrates are described in Chapter 5.

Flood-and-drain nutrient solution system

This hydroponic growing system had been in wide commercial use for many years, although it is not commonly used today. However, this method is still in use for hobby/home-type growing units. This system has also been called "ebb and flow." The growing system consists of a watertight rooting bed; rooting bed containing an inert rooting medium,

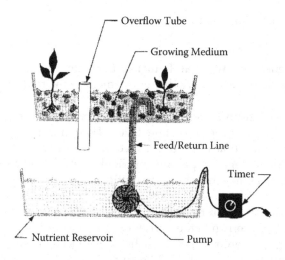

Figure 6.5 Flood-and-drain (ebb-and-flow) hydroponic growing system. Periodically, nutrient solution is pumped from the nutrient solution reservoir into the rooting medium, flooding the medium with nutrient solution, which is then allowed to drain back into the nutrient solution reservoir by gravity.

such as gravel, coarse sand, or volcanic rock; a nutrient solution sump (equal in volume to the growing bed or beds); an electrical pump for moving the nutrient solution from the sump to the growing bed or beds; and a piping system to accommodate the delivery of the nutrient solution from the sump to the growing bed(s) and its return.

Such a commercially designed system is shown in Figure 6.5. In order to have gravity return flow of nutrient solution from the growing bed(s) to the sump, the sump is placed below the growing bed(s). Being a "closed" system, the nutrient solution is recirculated until it is no longer usable; it is then discarded and replaced with fresh solution. Prior to each reuse, the nutrient solution is brought back to its original volume; tested for pH, EC, and possibly elemental content; and then adjusted accordingly. The nutrient solution may also require filtering and sterilization (see pp. 53–54) after each circulation through the rooting bed.

This hydroponic growing system was used by the US Army in World War II for the production of tomatoes and lettuce (Eastwood 1947). Following WWII, this system of hydroponic growing was put into commercial use by growers in several southern states in the United States, and elsewhere in outdoor hydroponic gardens growing primarily tomatoes. I have advised growers using this method of growing in both greenhouse and outdoor settings.

The disadvantages for this system are

- Susceptibility to root diseases
- Inefficient use of water and nutrient reagents
- Requirement for the periodic replacement of the rooting medium

A flood-and-drain system designed for greenhouse tomato production was marketed in the 1960s and 1970s. The sump held 2,000 gallons of nutrient solution that needed daily volume water adjustment as well as possible adjustments in pH and nutrient element makeup (based on an EC measurement). The nutrient solution required complete replacement about every 2 to 3 weeks—a considerably inefficient use of valuable water and reagents. With time, plant roots began to grow into the pipes that delivered and returned the nutrient solution to and from the growing bed(s) and sump, thereby restricting the flow. Once a diseased plant was introduced into the system, it would result in a total loss of the entire crop.

Cleanup frequently meant the removal and replacement of the gravel rooting medium. Another problem with this system was that because the rooting bed was in the ground, the sump and enclosed nutrient solution would have a temperature equal to that of the surrounding soil—meaning that, during most of the season, the nutrient solution would be colder than the ambient air temperature. This was an undesirable trait that would harm plants when the nutrient solution was dispensed into the growing medium.

For the hobby grower, the flood-and-drain system of growing is relatively easy to construct and operate on a small scale and gives reasonably good plant performance with a moderate level of care.

The timing schedule for flooding the growing bed(s) will depend on the atmospheric demand and stage of growth for the crop, as well as the water-holding capacity of the rooting medium. Normally, the composition of the nutrient solution is similar to the basic Hoagland/Arnon solution (see Table 4.6, p. 60) or some modification of it, depending on the crop and stage of growth.

Commercially, this system of hydroponic growing has proven to be difficult to manage and is very inefficient in its use of water and reagents—important reasons for its lack of use today.

Drip/Pass-Through Systems

There are two such growing systems: one using perlite or similar inorganic rooting medium (Morgan 2003b) in bags, pots, or buckets and the other using rockwool or coir blocks and slabs.

Inorganic rooting medium in bags or pots/buckets is in wide use for commercial production in which the plant is grown in a bag, pot, or bucket

Figure 6.6 BATO buckets (black for cool season use and beige for warm season use).

of inorganic medium, with perlite as the most common rooting medium (Gerhart and Gerhart 1992; Morgan 2003b). In one system, the bag used for shipping the perlite is laid on its side, small holes are cut along the bottom edge of the bag to allow excess nutrient solution to flow out, access holes are cut in the top of the bag for placement of a plant, and then a drip tube is placed on the edge of the access hole next to the plant. The plant may initially be seeded in a rockwool cube or other similar substance and then placed on an opening on the bag, with the drip line placed at the base of the plant. A pot or bucket, such as the BATO bucket filled with perlite or similar inorganic substance (see p. 90), can be used in place of the shipping bag. These systems, mostly using BATO buckets (Figure 6.6), are in wide use mainly for the production of tomato, cucumber, and pepper.

This method is normally operated as an "open" system (see p. 99), with the nutrient solution not recovered or reused. The amount delivered should be sufficient for a slight excess flow from the openings cut on the bottom edge of the bag or from openings in the bases of pots and buckets (the Bato bucket has a small reservoir in its base and an overflow siphon). Scheduling of the rate and timing of nutrient solution application is dependent on various factors, such as atmospheric demand, plant species, and stage of growth. During the growing period, the effluent from the growing vessel may be monitored for its pH and EC and adjustments made in the quantity of nutrient solution delivered, with the rooting medium being leached with water to remove accumulated salts. Also, an aliquot of solution can be drawn from the medium itself shortly after an irrigation to make the same measurements as made on an effluent sample.

Figure 6.7 Vertical PVC pipe with lettuce plants held in openings in the pipe with the nutrient solution being applied at the top of the pipe, flowing down through a rooting medium, or sprayed onto the roots from a holed-pipe placed in the middle of the PVC pipe.

At the end of the growing season, the perlite-containing vessel may be used one more time or discarded, which makes the system relatively easy to install and replace at a reasonable cost. The nutrient solution formula normally is based on the Hoagland/Arnon nutrient formula (see Table 4.1, p. 60) or some modification of it.

Various modifications of this system of growing have been made to accommodate different types of crops (Docauer, 2004). One example is a vertical hanging plastic-pipe column with lettuce plants placed in openings on the sides of the pipe—either containing a rooting medium, such as perlite, or the nutrient solution sprayed onto the plant roots functioning as an aeroponic system (DeKorne, 1992). For the nonaeroponic system, the nutrient solution is applied at the top of the pipe, usually through a dripper, and the solution passes down through the root mass or rooting medium and then out the bottom. A commercial application of this system is the tower garden (Figure 6.7). The same characteristics associated with the NFT technique apply to this system, for the composition of the nutrient solution is modified as it passes down through the rooting medium or root mass.

Figure 6.8 Column of interlocking Styrofoam buckets with a strawberry plant set in each open corner. The nutrient solution is applied at the top of the column of buckets, flowing down by gravity to be collected for recirculation or discarded.

Another system consists of a column of interlocking Styrofoam square pots or buckets in which plants are placed at the four corners of each pot (DeKorne 1998–1999; Devries 2003). The system is primarily designed for the growing of crop plants, such as lettuce and herbs, but can also be used for strawberries, flowers, and even small vine tomatoes (Figure 6.8). The nutrient solution outflow is normally collected and recirculated, either with (see p. 86) or without modification.

The advantage for these vertical systems is from the utilization of vertical space, thereby conserving lateral space if plants are grown in an enclosed shelter or greenhouse. The pipe or column of pots is usually rotated slowly to ensure uniform light exposure for the plants.

Subirrigation

This method of hydroponic growing has its basis in applying the "quality and balance" growing system developed by Geraldson (1963) combined with the findings of Asher and Edwards (1978a, 1978b) that low element content nutrient solutions when in "infinite" volume are sufficient to support normal plant growth.

The primary advantage of subirrigation is that all of the applied water and essential plant nutrient elements pass through the plant, so there is no loss of either to waste. The system can be operated by gravity flow, so there is no need for electrical power. My experience with long troughs reveals that there is a water flow mechanics issue, requiring the placement of a dispensing pipe in the bottom of long (greater than 4 feet) troughs.

By maintaining a constant level of nutrient solution (or water) in the base of the rooting vessel, roots will occupy that portion of the rooting medium where there is a balance between occupied water and air space. Root absorption of water and plant nutrient elements is controlled by root activity and therefore not influenced by periodic applications of either water or nutrient solution.

Today, two commercially marketed growing systems employ the subirrigation technique: the EarthBOX (www.EarthBOX.com) and the AutoPot (www.autopot.com). For the EarthBOX, the rooting container is filled with a soilless organic rooting mix containing sufficient essential plant nutrient elements to carry the plants through the growing season. The water level in the bottom of the EarthBOX is maintained by observing an indicator float in a stand pipe positioned in the corner of the EarthBOX. Water is added by hand through the float stand pipe when it is needed to maintain the proper water depth.

For the AutoPot system, each rooting vessel has its own float value that regulates the level of nutrient solution in the base of the pot, with the nutrient solution delivered to the value by gravity flow from a supply tank.

I have devised subirrigation rooting vessels, the *GroBox* and *GroTrough* (www.hydrogrosystems.com), where the level of nutrient solution in the base of the rooting vessel (either a box or trough) is maintained using a float valve housed within the rooting vessel or attached to it. The nutrient solution is delivered to the float valve by gravity flow from a supply tank. Photographs of the GroBox and GroTrough are shown in Figure 6.9 and Figure 6.10, respectively. A video describing both these growing systems can be found on the www.hydrogrosystems.com website. The best rooting medium has been found to be perlite with the nutrient solution being a one-third dilution of a modified Hoagland/Arnon formulation (Jones 2012b). Both growing systems are suitable for use outdoors as rainfall does not interfere with their functions.

Figure 6.9 A GroBox planted to tomato. The 9-gallon box on the right contains the nutrient solution; the box on the left has tomato plants rooted in perlite. A float valve is used to maintain a constant level of nutrient solution in the rooting box on the left. The float valve is housed in a PVC pipe located between the two 9-gallon boxes.

Figure 6.10 A GroTrough planted to tomato being grown in a greenhouse. The side PVC pipes provide support for the sides of the trough and can be used to anchor plant support stakes.

chapter seven

Hydroponic application factors

Introduction

When hydroponics was initially used commercially, only three crop species were being grown: tomato, herbs, and lettuce. Today a wide range of crop species (e.g., cucumber, pepper, strawberry, roses, potatoes, etc.) is being grown hydroponically. Even so, most commercially used hydroponic systems are still based on the requirements for growing either tomato or herbs and lettuce. In the early 1970s, the author visited a hydroponic greenhouse where the grower had successfully switched from growing tomatoes to chrysanthemum flower production in a gravel-sump flood-and-drain (sometimes referred to as ebb-and-flow) system using the same procedures as those given for tomato. This indicated to me that many different plant species can be successfully grown hydroponically, although the selected hydroponic system was not specifically designed for that plant species. Since that initial experience, it continues to be proven true. Today, a wide range of vegetables, flowers and even tree crops are being grown hydroponically using primarily two nutrient solution delivery techniques: flood-and-drain (see p. 108) and drip irrigation (see p. 110). The only exceptions would be for herbs and lettuce, where the Nutrient Film Technique (NFT) (see pp. 104–107) and the raft system (see p. 103) are the two systems preferred for use by growers.

An excellent example of what is possible hydroponically can be seen on display at the Kraft Exhibit in the Disney EPCOT Center, Orlando, Florida (Figure 7.1). Visitors taking the boat ride through the exhibit will see many different plant species being grown in various hydroponic configurations. A closer view of these growing systems, plus what experiments are being conducted but not on display, can be observed if the visitor takes the "behind the scenes" tour.

Hydroponic growing systems vary as to design, operational characteristics, and reliability, and they are generally more expensive and complex in their operating parameters than most other growing methods. Therefore, high-value cash crops (such as tomato) or specialty crops (such as herbs) are more frequently chosen for hydroponic production than are crops of lower cash value. Although initial costs may be high, hydroponics

Figure 7.1 View of the hydroponic display of the Land Exhibit at the EPCOT Center in Orlando, Florida. (Source: *Growing Edge* 8 (1): cover, Fall 1996.)

may be highly profitable as a method of crop production. Several of the major disadvantages of hydroponics include:

- High capital cost for most of the commonly used growing systems
- Frequent incidences of root disease
- High potential for nutrient element insufficiencies
- High operator knowledge and skill required

It should be remembered that hydroponics is not a panacea for success, no matter what crop is being grown what growing system is employed, or how skillful the grower may be. In general, the cultural requirements for a crop do not change even though a hydroponic growing technique is employed. In some instances, it may require greater skill on the part of the grower to be successful when a hydroponic method is used as well as greater attention to details.

Hydroponics does not invalidate the genetic character of plants, as plant growth and fruit production will not exceed what is genetically possible irrespective of the growing method used. Crop yields for plants grown in soil and in other soilless culture systems are comparable. Differences in yield favorable to hydroponics, when occurring, are due to efficient nutritional regulation gained by controlling water and plant nutrient element use as well as higher density planting. Since hydroponics does offer the ability to control the supply of water and the essential plant nutrient elements to plant roots, a continuous optimum supply can in turn enhance plant performance. Greenhouse-grown crops, such as tomato, cucumber, pepper, and lettuce, can be grown over a longer time period than that possible outdoors or in minimum climate-controlled structures. Therefore, yield comparisons for that hydroponically grown in environmentally controlled structures versus soil field grown can be misleading because, when compared on equal environmental terms, yields are usually similar.

Schoenstein (2001) states,

> In addition to the organic angle, controlled-environment greenhouse agriculture allows farmers to reach more niche markets due to their ability to extend a crop into a much longer season than that available to outdoor growers. With the rise in the huge, large-scale greenhouses in North America, the value of conventional off-season produce is decreasing while the value for organic crops remains high.

The public's increasing interest in wanting to purchase "organically grown" produce may significantly impact the future of hydroponics. Being just pesticide/herbicide free is no longer the single factor that attracts the environmentally sensitive consumer who is looking for food products that are organically grown. However, the switch from inorganic to organic hydroponics will require the development of suitable growing media and nutrient solution formulations that will qualify as being organic. Schoenstein (2001) describes a greenhouse operation that is producing lettuce and herbs using an organic NFT growing system.

When growing conditions are such that no other system of growing is suitable—for example, due to poor soil conditions, in extreme climatic regions (deserts, arctic regions), and in outer space and roof-top gardening (Wilson 2002a)—hydroponic production may be the only option.

Progressive developments

The initial hydroponic growing system was the standing aerated nutrient solution method (see p. 99), a method considered unsuitable for commercial plant production. However, Cunningham (1997) describes the use of this technique (which he identifies as the updated Gericke system; Figure 7.2) for the growing of green bean, tomato, and zucchini squash—a system that does not require electrical power and is fairly simple to use. Kratky (1996) describes the general principles and concepts of a noncirculating growing system for hydroponically growing lettuce, tomato, and European cucumbers.

Wilcox (1983) wrote an extensive review of those hydroponic systems in use throughout the world at that time, water or solution culture, sand culture, aggregate culture, and the nutrient film layout. For commercial applications, the flood-and-drain method (Fischer et al. 1990) was the initial hydroponic growing system, closely followed by the gravity flow bed system. There are other techniques that have specific applications, such as the raft system for lettuce production (Spillane 2001; Morgan 2002c) and aeroponics (see p. 108) (Nichols 2002; Wilson 2002b).

In 1979, the Nutrient Film Technique (NFT) developed by Cooper (1979b) was hailed as a revolutionary step forward that would alter the method of hydroponic growing for all crops. But this did not prove to be true (see pp. 104–107). Today, the NFT method is primarily used for

Figure 7.2 Updated Gericke system for growing garden vegetables. Plants are suspended over a nutrient solution reservoir with an air gap between the base of the plants and suspended roots. (Source: Coene, T., 1997, *Growing Edge* 8(4):34–40.)

the hydroponic production of lettuce and herbs. Morgan (1999b) describes the various designs for gullies and channels for use in NFT applications. Smith (2004) also gives instructions for the design and construction of NFT gullies.

With the introduction of drip irrigation, water and/or a nutrient solution could be dispensed at a specific point and precise volume. Using this method, growers can grow hydroponically in containers of inert media, such as in perlite (Day 1991), bags (Bauerle 1984), BATO buckets (see Figure 6.6 in Chapter 6), or rockwool blocks and slabs (Figures 5.1 and 5.2 in Chapter 5). Today, this is the primary technique of choice for the growing of tomato, cucumber, and pepper.

Organic substances are also being used as rooting media, such as composted milled pine bark (Pokorny 1979) and coconut fiber (Morgan 1999a), which have an environmental advantage over perlite and rockwool (Spillane 2002) since they are biodegradable. Morgan (2003b) describes the properties and use of 18 growing media substrates: rockwool and stonewool, vermiculite, perlite, coconut fiber, peat, composted bark, pea gravel, metal chips, sand, expanded clay, sawdust, pumice, scoria, polyurethane grow slabs, rice hulls, sphagnum moss, vermicast, and compost. She then matches substrate characteristics with a particular hydroponic growing method (Table 7.1).

Nutrient solution formulations and their use

A detailed discussion on nutrient solutions, their formulation, and their use is given in Chapter 4.

Table 7.1 Substrate Characteristics Matched to a Particular Hydroponic Growing Method

Hydroponic method	Substrate characteristics
Ebb and flow	Must be reasonably heavy so that it will not float away; drain reasonably well, although hold some moisture; materials such as expanded clay, gravel, coarse sand, pumice, or rock-like material
Drip irrigation systems	Must hold a reasonable amount of moisture; high percentage of air filled pores
Warm climates	Heavy media types that hold more water and are slow to heat up; materials such as coconut fiber, ground bark, rockwool, or stonewool
Cooler climates	Prevention of continually cold wet root system important; freer draining; materials such as perlite, pumice, sand, and expanded clay

Source: Morgan, L., 2003b, *Growing Edge* 15(2):54–66.

The literature is filled with specific formulations that have been pre-scribed for a particular plant species and/or hydroponic method. For example, in *The Growing Edge* magazine in issues published between 1989 and 2002, Jones and Gibson (2003) found 19 articles on the formulation and use of nutrient solutions and some 32 specific nutrient solution formulas recommended for various plant species. Plant species requirements could be a major factor that would specify the need for a particular formulation as well as a hydroponic growing method. But in most instances, it is the use factors, such as frequency of application and volume per application, that are sketchy and leave the reader confused as to when and how the nutri-ent solution is to be dispensed to the plant. Therefore, these formulation recommendations require careful evaluation before their acceptance and use. In addition to the type of hydroponic growing system and plant spe-cies being grown, the environmental conditions are equally influencing factors that could affect formulation/use recommendation modification.

It is the author's opinion that there is no justification for most spe-cific modifications based on plant species and hydroponic method. It is also probable that there are several basic formulations that are suitable for wide use, such as modifications of the Hoagland/Arnon formulations (Hoagland and Arnon 1950, p. 60). Those wanting to make their nutri-ent solution from scratch will find the instructions by Musgrave (2001) helpful, covering the topics of rule of conversion, determining elemental percentages and the rest of the conversion theory.

Cultivar/variety availability and selection

Plant cultivars/varieties identified as best suited for only hydroponic growing do not exist. Breeding and selection are based on developing plants best suited to a specific environment, such as day length and light intensity, or plant characteristics, such as disease resistance, drought and/ or heat tolerance, fruiting habit, and fruit characteristics.

Much has been written about "genetically modified" varieties—modi-fied in order to obtain a particular characteristic, which is a topic that has stirred much comment and controversy. Most of the breeding work has been focused on crop plants commonly grown and those that would be classed as having a "high cash value," such as tomato. Much of the breeding work has been focused on cultivar development where the need is greatest. For example, greenhouse tomato cultivar breeding and selec-tion have been for adaptation to low-light, low-temperature conditions, while little attention has been given to cultivars that would have high-light, high-temperature tolerance. In addition, a fruit's physical appear-ance, color, firmness to withstand rough handling, storage quality, etc., are some of the qualities being bred into the newly released cultivars.

Cultivar/variety selection is a major decision that the grower faces as a mis-selection can lead to poor plant performance and low fruit quality.

Recently there has been interest in "heirloom" varieties, which are those that have a significant history of acceptance and use. However, many of these varieties do not perform well in some types of growing systems, whether hydroponic or not. Also, many heirloom varieties lack specific disease resistance, one of the major focuses in the introduction of new varieties as well as sensitivity to changing environmental conditions.

Constancy

Constancy of growing conditions leads to high yields and quality product production. It is not possible to control every aspect of the environment precisely (the amount of radiation received, for example) or to control the continuous cycling of atmospheric conditions within the greenhouse or outdoors adequately. In most greenhouses, it is possible to minimize the cycling of the air temperature, CO_2 content, humidity, etc., by the use of computer-driven control devices (Lubkeman 1998, 1999; Nederhoff 2001). Growth chamber experiments have demonstrated the effect that precise control of the aerial environment can have on the growth and development of plants. Therefore, a greenhouse system must be designed to mimic what is possible in a growth chamber if those environmental conditions required for high growth are to be obtained and maintained.

For most of the commonly used hydroponic growing systems, the cycling of water and nutrient element availability are not easily controlled. As a nutrient solution is introduced into the growing medium, three things occur. Plant roots absorb the water and nutrient elements in the nutrient solution at varying rates (Bugbee 1995), water and nutrient elements that are not absorbed begin to accumulate in the rooting medium (Jones and Gibson 2002), and some of the applied water and nutrient elements pass through the rooting vessel to be discarded to waste or collected for recirculation. The result is a continuously varying rooting environment that can adversely affect plant growth. This is one of the influencing factors that is not being adequately addressed by those engaged in hydroponic system research and development (www.hydrogrosystems.com).

Grower skill and competence

As with any plant-growing venture, the skill of the grower can mean the difference between success and failure irrespective of the operational quality characteristics of the growing system. Some attribute this to a *green thumb* ability that some individuals seem to have—that sense to know what to do and when that leads to maximum plant performance. The author has visited many greenhouses and, just by looking around, it does not take long to assess the skill and ability of the grower to manage the crop and greenhouse facility. For example, just the physical appearance of the crop, such as color, growth habit, and freedom from insect and disease infestations, can be a good indicator of grower skill. Also, answers to the following questions provide an assessment of grower attention to detail:

- What has been the timeliness of applied cultural practices?
- What is the general condition of the greenhouse structure, inside and out, its cleanliness, and the condition and efficiency of the heating, cooling, and air distribution systems?

These are some of the observable things that can be used to determine the competence of the grower and workers. Smith (2001a, 2001b), for example, gives advice on what an NFT tomato grower needs to do when the crop is in full production to sustain fruit yield; this advice can be applied to any hydroponic grower evaluation (Table 7.2). What have been the past training experiences? Much can be learned from practical experience and/ or hands-on training under the tutelage of a knowledgeable instructor.

It has been my experience that most hydroponic growing system failures occur due to several key factors. During the 1970s, I witnessed the collapse of a hydroponic industry in the state of Georgia. It occurred due to two primary factors: (1) the poor design and inefficiencies of both the greenhouse and hydroponic growing system, and (2) the lack of experience and professional skill required to manage the greenhouse/hydroponic system successfully on the part of the managers/workers. At about the same time, I observed the success of a small group of tomato greenhouse growers who were trained and guided by a skilled, experienced

Table 7.2 Instructions on Procedures for Managing an NFT Tomato Crop

Immediately apply a fungicide on plant wounds, particularly if *Botrytis* is a commonly occurring disease

Remove laterals and then immediately lower the plant

Before fruit harvesting, remove all senile leaves (will reduce potential for disease and other plant problems)

Keep the nutrient solution barrels and measure and correct the pH if outside the pH range of 6.0 to 6.5

Cut fruit using sharp pruning scissors ; keep the calyx on the fruit to maintain fruit quality, and refrigerate at a temperature just above 55°F (13°C)

In cold weather conditions, keep the rooting medium and nutrient solution at a temperature above 59°F (15°C), and the air temperature within 70°F to 77°F (21°C to 25°C)

Sources: Smith, B., 2002b, *Growing Edge* 13(4):75–79; Smith, B., 2002, *Growing Edge* 13(5):79–82.

Figure 7.3 Miniature greenhouse structure with the base containing the nutrient solution that is periodically pumped into the gravel bed for rooting plants.

professional in southeast Georgia. When that individual left to take another position, many of the growers he trained and guided closed their greenhouses, fearing that trying to continue without his guidance would eventually lead to failure.

An entrepreneur in central Florida established a successful hydroponic business growing tomatoes in semienclosed structures using a flood-and-drain gravel medium system (Figure 7.3). When the root disease *Phytium aphanidermutum* entered the rooting pea-gravel medium, he was unprepared to deal with this infestation, and his business failed in less than 6 months.

Grower success is hinged on procedural factors other than the individual's innate skill. Having the input of professionals in all aspects of the growing method can significantly contribute to success. However, no amount of grower skill and professional guidance can overcome the effects of a poorly designed greenhouse/hydroponic growing system.

Factors for success

There is no one hydroponic technique that is applicable to every plant growing situation in terms of method (flood-and-drain, NFT, media systems using drip irrigation) and control of the supply of water and nutrient solution (their formulations). A critical factor is where the hydroponic system is being put into use, whether in a greenhouse with its wide variance in design and function, in a controlled environmental chamber, or outdoors. The physical location of a greenhouse or outdoor site in terms of specific location at a particular site and/or in regions with varying weather conditions (high and low temperatures, and high and low light intensity and duration) will govern what will be required to be successful.

The wealth of information on hydroponics that is available can easily lead to making wrong choices in the design of the growing system and operating procedures. A common error is to adopt a growing system and/or set of operating procedures that are only applicable to a particular environmental situation. For example, what would be required under low temperature and light would not necessarily apply equally under high temperature and light. What would be required for a crop under declining light, from summer into the winter months, would not apply under increasing light conditions from winter into the summer months, being identified as either a fall or spring crop.

The skill of the grower and experience with a particular growing method may not be easily transferable to an inexperienced grower. I have visited with growers whose success could be directly related to their innate skill, a sense that directs what to do and when to do it (the *green thumb* phenomenon). It is always better to be proactive than reactive to changing conditions of the crop or growing system. This is particularly true when environmental conditions change and alter the water and nutrient element requirements of the plants, or when there is need for shade or increased light. Failure to anticipate significant weather events, such as unexpected low or high air temperatures, snowfall, or high winds, can result in damage to the greenhouse structure as well as to the status of an enclosed crop.

When dealing with plant nutrition and pest problems, relying on knowledgeable professionals for identification and recommendations is essential. Monitoring and periodic testing are required to ensure that the crop is being sufficiently maintained nutritionally (see Appendix D).

Unfortunately, not even the best designed growing system or greenhouse structure will initially perform well—even in the hands of a skilled grower. It may take a "shakedown" period to make the total system work efficiently and the growing plants to perform to expectations.

What might function well under one set of environmental conditions may not work under another. This was the experience of four growers located in the southeastern United States, who initially produced high-yielding and -quality fruit, but then experienced low yields and quality in the following years (Jones and Gibson 2001). The ability to find the source of a problem and then to adjust to or correct it makes for success as well as minimizing losses.

Record keeping is essential if a grower is to continue to produce high-yielding crops of high quality. Environmental conditions should be recorded daily. The dates when major events occurred and the changing status of the plants should also be recorded. Accurate yield records plus quality evaluations are essential. There was a greenhouse tomato grower who kept accurate weekly fruit production records and then correlated obtained yield with the amount of weekly sunshine, data that were gathered from the local weather station. The highest correlation found between fruit yield and weekly sunshine was obtained 2 or 3 weeks prior, demonstrating that the effect of light intensity conditions did not appear in terms of fruit yield until several weeks later. In addition, if yields are compared with daily and/or weekly growing conditions, these data can be used to guide the grower when decisions must be made in the future (Nederhoff 2001).

Crop and cultivar selection should be based on adaptability to the growing system and environmental conditions as well as the marketability of the harvested crop. A common error is to produce a crop that does not meet or exceed the market requirements and/or is not of sufficient quality for consumer acceptance. A grower may be very successful in producing a crop but may not be able to market it adequately. As mentioned earlier (see pp. 124–125), one of the reasons a group of southeastern Georgia tomato greenhouse growers were successful was not only that each grower had his own local market, but also all the growers were able to pool their surplus fruit, which was taken to a centralized market for sale in a large nearby city.

For most of the commonly used hydroponic growing systems, the cycling of water and nutrient element availability are not easily controlled. As a nutrient solution is introduced into the growing medium, three things occur: Plant roots absorb the water and nutrient elements in the nutrient solution at varying rates (Bugbee 1995), water and nutrient elements not absorbed begin to accumulate in the rooting medium (Jones and Gibson 2002), and some of the applied water and nutrient elements leach from the rooting vessel. The result is a continuously varying rooting environment that can adversely affect plant growth. This is one of the influencing factors not being adequately addressed by those engaged in hydroponic system research and development. Geraldson (1963, 1982) has addressed this problem in his research on the effect of quantity and balance of the nutrient elements on the growth of field-grown staked tomatoes. This basic concept was used by Jones and Gibson (2002) in their development of the *Aqua-Nutrient* growing method and is the basis for a commercial product called the "EarthBox" (Figure 7.4).

There is much yet to be learned about how best to grow plants hydroponically. There have not been any significant breakthroughs in the last several decades. Most of the hydroponic growing systems in use today were developed in past years. What the future holds for new developments is uncertain since few are engaged in developmental research on hydroponic methods.

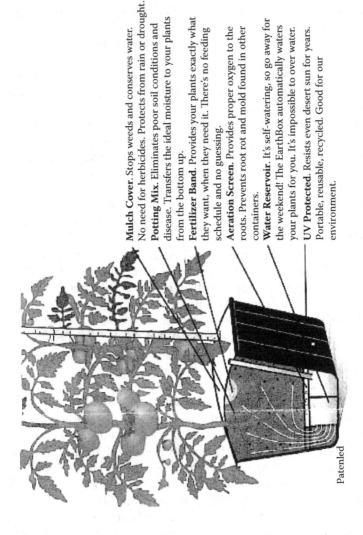

Mulch Cover. Stops weeds and conserves water. No need for herbicides. Protects from rain or drought.

Potting Mix. Eliminates poor soil conditions and disease. Transfers the ideal moisture to your plants from the bottom up.

Fertilizer Band. Provides your plants exactly what they want, when they need it. There's no feeding schedule and no guessing.

Aeration Screen. Provides proper oxygen to the roots. Prevents root rot and mold found in other containers.

Water Reservoir. It's self-watering, so go away for the weekend! The EarthBox automatically waters your plants for you. It's impossible to over water.

UV Protected. Resists even desert sun for years. Portable, reusable, recycled. Good for our environment.

Patented

Figure 7.4 EarthBox. (EarthBox, PO Box 420, St. Petersburg, FL 33731-0420.)

Controlled-environment agriculture

Today, most hydroponic growing is being conducted in an environmentally controlled climate, such as a greenhouse or enclosed chamber using artificial lights. Hydroponics is not generally considered a method of growing in the open (outdoor) environment. However, it is interesting to note that open-environment hydroponic systems were in wide use during WWII, when vegetables were grown hydroponically to provide fresh produce for troops operating in the Pacific campaign areas. After WWII, in 1950, I visited a number of hydroponic farms in South Florida; the crop grown was tomato in the open environment in ebb-and-flow gravel beds. Some of the early literature on hydroponics in the 1950s and 1960s discussed methods of growing in the open field (Eastwood 1947; Schwarz 2003). One wonders whether the future of hydroponics as a major method for crop production lies in other than open field settings as it was in its initial years of application.

Today, much of the current literature on hydroponics discusses this topic as a form of controlled-environment agriculture (CEA). Therefore, much of the success associated with hydroponics may be more related to environmental control advances than to those associated with the hydroponic method being employed (see http://ag.arizona.edu/ceac).

Hydroponic growing systems vary as to design, operational characteristics, and reliability, and they are generally more expensive and complex to operate than most other growing methods. Therefore, high-value cash crops (such as tomato) or specialty crops (such as herbs) are more frequently chosen for hydroponic production than are crops of lower cash value. Although initial costs may be high, hydroponics can be highly profitable as a method of crop production. Several of the major disadvantages of hydroponics are the high capital cost for most of the commonly used growing systems, frequent incidences of root disease, and the potential for nutrient element insufficiencies. However, these factors are being addressed and advances made to solve the problems of cost and insufficiencies related to the hydroponic method of growing.

It should be remembered that hydroponics is not a panacea for success, no matter what crop is being grown or what growing system is employed. In general, the cultural requirements for a crop do not change even though a hydroponic growing technique is employed. In some instances, it may require greater skill on the part of the grower to be successful when a hydroponic method is used.

Hydroponics does not invalidate the genetic character of plants; plant growth and fruit production will not exceed what is genetically possible irrespective of the growing method used. Hydroponics does offer the ability to control the supply of water and the essential nutrient elements to plant roots, thereby ensuring a continuous optimum supply, which can in turn enhance plant performance. Greenhouse-grown crops, such as tomato, cucumber, pepper, and lettuce, can be grown over a longer time period than that possible in the outdoors. Therefore, yield comparisons for hydroponically grown versus soil field grown can be misleading, for if equably compared, yields are similar.

Outdoor hydroponics

Outdoor application is the form of hydroponics least studied and applied, yet it has significant potential. The advantages of growing outdoors hydroponically are many; the primary ones are the ability to control the use of water and plant nutrient elements and the avoidance of soil-related challenges, such as poor soil physical and chemical properties as well as weeds and soil-borne diseases.

I have had good success growing a variety of garden vegetables (tomato, pepper, lettuce, green beans, strawberry, melons, sweet corn, okra) outdoors hydroponically in a system in which a depth of nutrient solution is maintained in the bottom of watertight vessels, boxes, and troughs as shown in Figures 6.5 and 6.6 in Chapter 6 (see pp. 114–116). A detailed description of the basic principle of operation for this method can be found at the website www.hydrogrosystems.com.

From their extensive knowledge and experience Bradley and Tabares (2000a, 2000b, 2000c, 2000d) and Bradley (2003) describe how simplified hydroponic growing systems are being built in developing countries not only to alleviate hunger but also to create small business ventures. Included are easy-to-follow instructions and operational principles for growing systems that would be of value to anyone interested in getting started in hydroponics.

The personal experience of Ray Schneider—an energetic hobbyist who first began indoors (Schneider 1998) and then went outdoors (Schneider 2000, 2002, 2003, 2004) with his NFT hydroponic system—is an example of the successes and pitfalls that can occur. In an article by Schneider and Ericson (2001), the learning experiences of Ericson, who used 6-inch sewer pipes as the growing vessel to grow hydroponic lettuce, bell pepper, tomatoes, cabbage, parsley, and herbs, are described. Christian (1997) describes an NFT lettuce-growing system that was designed based on what had been done elsewhere, and how crop protection devices were designed and used to deal with weather extremes that occurred from time to time. Kinro (2003) describes how Larry Yamamoto in Honolulu, Hawaii, turned a hobby into a career growing hydroponic lettuce using a simple raft floating system.

Home gardener/hobby hydroponic grower

Hydroponics offers the home gardener and hobbyist a challenge that some have undertaken. Most have devised their own hydroponic growing systems based on information found in books, manuals, magazine articles, and on the Internet. In a four-part series, Smith (2001a, 2001b, 2001c, 2001d) describes how to design and build your own hydroponic system. He states, "My introductory hydroponic series delved into the fundamentals of what makes hydroponics tick. We discussed the quality of your water supply, the different types of systems, and the hydroponic nutrition required by your plants." "Hydroponics for the Rest of Us" is the title of an article that details various hydroponic growing systems (passive—wick system and active—flood-and-drain, top feed, NFT) and their operating requirements. The two recommended for the home gardener are the flood-and-drain and top feed hydroponic systems because they "lend themselves well to home design and construction without sacrificing durability and efficiency" (Van Patten 1992). Van Patten (1992) divides systems into two additional categories—recovery or nonrecovery (recirculation or discard, respectively)—of the nutrient solution.

Coene (1997) provides basic information on soilless gardening, focusing on media- and water-based culture systems, nutrient solutions, artificial lighting, and pest control; she then describes how to build a drip system growing vessel (Figure 7.5). Similarly, Creaser (1996–1997) gives instructions for the construction of a drip system growing tray that he has been using to grow an array of vegetables and house plants (Figure 7.6).

Peckenpaugh (2002a) recognized the need of hobby growers to have a reliable source of information on hydroponic techniques and procedures. He describes the design and operation for four hydroponic systems (NFT, floating raft, flood-and-drain, and drip) and the formulation and use of nutrient solutions including organic, identifying those crops (cucumber, lettuce, pepper, strawberry, and tomato) most commonly grown plus how to deal with insects and diseases. For one who wants just to experiment with small growing systems, Peckenpaugh (2002a) describes hydroponic growing techniques that "can be built by anyone with the time and patience to go through the process." In his article, he describes three different growing systems, passively wicking pot, styrofoam cooler grower, and Dutch pot dripper; lists the materials needed to construct each of the systems (Table 7.3); and describes how to assemble and operate each.

For those wanting to construct their own drip irrigation hydroponic growing system, Peckenpaugh (2003a) lists the following items required: growing container, drip irrigation lines, drip emitters, nutrient reservoir, submersible pump, nutrient return line, growing media (expanded clay), and timer. He states that "drip irrigation approaches the pinnacle of growing sophistication due to its highly economical use of water and precise

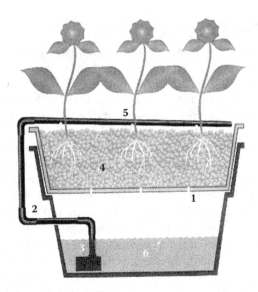

Figure 7.5 Drip system hydroponic growing vessel for growing vegetables. 1: Small holes in the base of the growing vessel allow applied nutrient solution to flow back into the nutrient solution reservoir (6); 2: nutrient solution delivery line; 3: pump; 4: pea gravel rooting medium; 5: drip irrigation line. (Source: Coene, T., 1997, *Growing Edge* 8(4):34–40.)

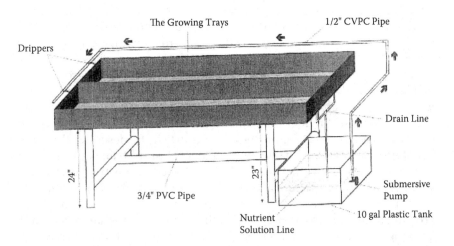

Figure 7.6 Drip irrigation hydroponic growing tray for growing vegetables. (Source: Creaser, G., 1997, *Growing Edge* 8(3):86–93.)

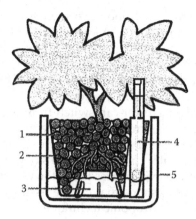

Figure 7.7 Hydroculture system for growing plants in a small pot. 1: Clay; 2: culture pot insert; 3: nutrient solution; 4: water level indicator; 5: outer container. A constant level of nutrient solution is maintained in the bottom of the outer container, with nutrient solution added through the water level indicator opening. (Source: Angus, J., 1995–1996, *Growing Edge* 7(2):48–55.)

application of nutrients to the plant's root zone." Expanded clay is the growing medium since it "holds some moisture and nutrient for plant use after the irrigation cycle but doesn't get soggy or overly wet." In addition, expanded clay can be easily sterilized after use by baking at 400°F (204°C) for an hour. In a follow-up article, Peckenpaugh (2003b) describes his success in using his homemade hydroponic growing system.

Alexander and Coene (1995–1996) focused on those hydroponic systems that would attract the cost-conscious grower who does not want to make a significant investment in equipment. A simple passive hydroponic system, described by Christensen (1994b), may be a good place to begin one's initial venture into hydroponics; it is a spin-off of an earlier

Table 7.3 Materials Needed to Construct Your Own Hydroponic Growing System

Growing system	Materials
Passively wicking pot	Two pots or buckets, wicking material, wire (optional), utility knife or drill, growing media
Styrofoam cooler grower	Styrofoam cooler, polythylene or garbage bag, duct tape, propagation cubes, utility knife, aquarium pump, tubing, and air stone (optional)
Dutch pot dripper	Pots, nutrient solution reservoir, submersible pump, runoff collection trough (optional), fine mesh screening, drip irrigation tubing, emitters, and support stakes

Source: Peckenpaugh, D. J., 2002b, *Growing Edge* 13(4):81–83.

described noncirculating hydroponic system (Christensen 1994a). The recent book by Roberto (2001) provides a "guide to build and operate indoor and outdoor hydroponic gardens, including detailed instructions and step-by-step plans." Resh (2003) has a book on hobby hydroponics "to provide the reader with information on the basics of hydroponics that can be applied to a small-scale or hobby setup."

Even the houseplant grower can switch from in soil to hydroponics. Angus (1995–1996) gives instructions for how a hydroculture system consisting of five basic parts—clay pellets, nutrients, water level indicator, culture pot insert, and outer container—works (Figure 7.7).

All these articles and experiences tell of the wide range of opportunities as well as growing systems that can be used by those interested in experimenting with the hydroponic technique.

chapter eight

Educational role

Introduction

In 1993, Brooke and Silberstein (1993) wrote,

> Increasingly, hydroponics, long a tool of university researchers, is finding a place in elementary and secondary education. It offers students great opportunities to learn from their successes and from their failures. On the way, they learn about making observations, recording and interpreting data, and the need for control in scientific research.

Peterson Middle School, located in Silicon Valley, began a hydroponic project in 1992 by designing a simulated space capsule (ASTRO 1) to house various hydroponic growing systems (Figure 8.1). "Tending the plants and monitoring their progress teaches responsibility and a respect for living organisms that no textbook biology lesson could convey" was the comment made by the two lead teachers (Brooke and Silberstein 1993). The lettuce and tomatoes produced were served in the school's cafeteria. The ASTRO 1 project has led to teacher presentations at hydroponic camps and teacher conferences (Silberstein 1995), as well as workshops for teachers held mainly in California with the assistance of the Hydroponic Society of America (HSA) (Silberstein and Spoelstra-Pepper 1999).

Recognizing the need to "know where to look to find valuable information for all grade levels, the right lesson plan, that important grant, and intuitively designed equipment," Hankinson (2000) developed "a hydroponic lesson plan" to guide the teacher on topics and experiments for instructing students in hydroponic studies, including the following topics: plant nutrition basics, preparing the plants, containers, aeration, and nutrient solution. The author then described how to conduct hydroponic experiments to produce nutrient element deficiency symptoms in test plants. Similarly, Hershey (1994) gives instructions for the teacher on how inexpensive equipment can be used to conduct hydroponic experiments. Those interested in a wider range of plant biology science projects will find the book by Hershey (1995) helpful. For the more advanced student, Morgan (2002a) provides sources of instructional material on all

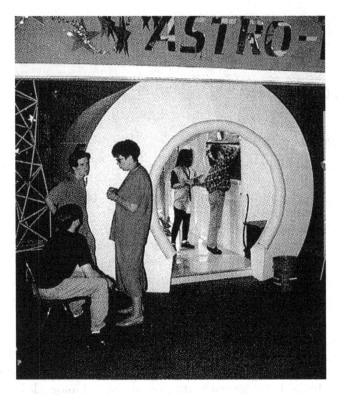

Figure 8.1 Peterson Middle School hydroponic simulated space capsule, ASTRO 1. The crops produced are served in the school cafeteria. (Source: Brooke and Siberstein, *Growing Edge* 5(1): 20–23, 1993.)

aspects of the hydroponic growing technique, nutrient solution formulations and their chemical characteristics, and systems of plant production Morgan (2002a) describes experiments that can be conducted in the classroom (Table 8.1).

Demonstration project

A common school science fair project demonstrates some form of hydroponic plant growing, and the information presented here can be helpful to a student contemplating such a project. The student wishing only to grow plants hydroponically to demonstrate the technique can use the Knopp nutrient solution formulation (see p. 19). The best rooting medium is perlite.

The hydroponic procedures that a student can follow in order to generate nutrient element deficiency symptoms and monitor their effects on plant growth and development are described. With easily obtainable

Table 8.1 Procedures and Objectives for Students Conducting Hydroponic Experiments

1.	Use the scientific method whenever cause and effect are documented.
2.	Include the objective of the experiment and a hypothesis.
3.	Make a comprehensive list of required materials.
4.	Start experiments with plants of a similar size and stage of development and clearly label each plant.
5.	Keep track of all growing variables, such as daily temperatures, changes in nutrient electrical conductivity (EC) and pH, amount of solution used, and any pest and disease problems encountered.
6.	Record all observations.
7.	Discuss results and ask questions; results that show that nothing happened are still valid and worth reporting and discussing.
8.	Outline how the experiment was undertaken in detail, using photographs and diagrams.
9.	List all sources consulted, including any expert advice, books, magazine articles, websites, and so on; such documentation shows that a well-researched experiment has been conducted.

Source: Morgan, L., 2002a, *Growing Edge*, 13(6): 56–70.

items and properly prepared nutrient solutions, the student should be able to undertake such a science project and obtain good results in about 6 to 8 weeks. The references in this section provide specific information that will be helpful to a student with this project.

Required items

The items required for this project are as follows:

1. One-liter plastic beverage bottles
2. Horticultural perlite (available at most garden centers)
3. 6 × 6 in. plastic refrigerator storage boxes with lids
4. 50 mL graduated cylinder (available from most chemical apparatus supply houses)
5. Green bean seeds (bush beans recommended)
6. Pure water
7. Nutrient solution reagents (obtainable from most chemical supply houses or hydroponic suppliers):

Reagent	Formula
Major element reagents	
Calcium nitrate	$Ca(NO_3)_2 \cdot 4H_2O$
Calcium sulfate	$CaSO_4 \cdot 2H_2O$

Potassium nitrate	KNO_3
Potassium sulfate	K_2SO_4
Monopotassium phosphate	KH_2PO_4
Magnesium sulfate	$MgSO_4 \cdot 7H_2O$
Magnesium nitrate	$Mg(NO_3)_2 \cdot 6H_2O$

Micronutrient reagents

Boric acid	H_3BO_3
Copper sulfate	$CuSO_4 \cdot 5H_2O$
Manganese chloride	$MnCl_2 \cdot 4H_2O$
Manganese sulfate	$MnSO_4 \cdot H_2O$
Molybdic acid	$H_2MoO_4 \cdot H_2O$
Zinc sulfate	$ZnSO_4 \cdot H_2O$
Iron chelate	FeDTPA

Growing requirements

Light

Plant growth is best when plants are exposed to full sunlight for at least 8 hours each day. Placing plants by a window, even one that is well lit, is not sufficient for best growth. Use of lights to extend the exposure time is not an adequate substitute for natural sunlight. Slow growth and development, usually seen as spindly looking plants, are signs of inadequate light. If a nutrient element deficiency symptom is to be developed, plant exposure to full sunlight is essential.

Plant species selection

For best results in a reasonable length of time, a plant species that grows rapidly and is responsive to its environment should be selected. Experience has shown that green bean is probably the best plant species, and corn is second best. Although other plant species, such as radish and lettuce, are faster growing, the larger plant size of the green bean and corn plants makes them the best choices.

Temperature

Plants grow best when the air temperature is maintained between 75°F and 85°F (24°C and 30°C). Air temperatures above or below these limits are not conducive to normal plant growth and development.

Moisture

Plants that are cycled through periods of adequate and then inadequate water supply will develop abnormal growth appearances due to that stress. Therefore, plants must have access to an adequate supply of water at all times. However, overwatering is as detrimental to plant growth and development as inadequate watering. Frequent small doses of water added to the rooting medium are better than infrequent heavy doses. The growing technique described in this section will maintain an adequate water supply for the plants at all times.

Pest control

Insects and disease problems can be avoided by keeping the growing area clean at all times and free from potential sources of infestation. Although neighboring plants may be free from visible pests, it is wise to conduct the experiments given in this section free of the presence of other plants that are not part of the study.

Procedure

1. Remove the top of a 1-liter plastic soft drink bottle by cutting around the bottle at the upper label level (Figure 8.2). The number of bottles needed will depend on which experiments will be conducted. Only

Figure 8.2 One-liter soft drink bottle set in the refrigerator box.

one bottle per treatment is needed, although duplicates will ensure that there will be a backup treatment bottle if one bottle is lost. Also, one bottle plus its backup will be needed as the check (that without a treatment change).

2. Drill a 1/4 in. diameter hole in the center of the bottom of the bottle.
3. On the inside of the bottle, cover the hole in the bottom of the bottle with plastic mesh. The plastic mesh will prevent the loss of perlite from the opening in the bottle.
4. Fill the bottle with horticultural-grade perlite all the way to the top.
5. Using the prepared nutrient solution, leach the perlite until the nutrient solution freely flows from the hole in the bottom of the bottle. The plastic mesh, if properly in place, will keep the perlite from being lost from the bottom of the bottle.
6. Place two green bean seeds about an inch deep into the moist perlite. It may be necessary to add a small amount of water (half a cup) daily to the top of the bottle to keep the perlite moist until the seeds germinate and the cotyledons appear.
7. Place the bottle into a small plastic refrigerator box and fill the box to a depth of about 2 in. with nutrient solution. With a black marker pen, put a scribe mark at the nutrient solution level on the side of the refrigerator box. When adding nutrient solution, always fill to that mark. Cut an opening in the box lid sufficient to accommodate the bottle. Place the refrigerator lid on the box and snap down tight (Figure 8.2). Keeping the lid in place will prevent evaporation of the nutrient solution. The nutrient solution in the box will also fill the bottle with nutrient solution at that same level.
8. Place the bottle in its box in full sunlight. Add nutrient solution when needed (usually every day) to maintain the level in the box at the scribe mark using a 50 mL graduated cylinder so that water use can be monitored.
9. When the seeds germinate, remove one of the seedlings to leave just one plant per bottle.
10. When the plants reach the two-leaf stage, begin the treatments.

Nutrient element deficiency experiments

Visual nutrient deficiency symptoms for the major elements (Ca, Mg, N, P, and K) are fairly easy to develop using the technique that is to be described. Creating a deficiency of any one of the micronutrients (B, Cu, Fe, Mn, and Zn) is more challenging and difficult to achieve. The reason is that the major elements are required in substantial quantities by plants, whereas the micronutrients are not. It is quite difficult to deplete the growing medium and the nutrient solution of trace quantities of the micronutrient elements in order to create a deficient condition. In addition, there

Table 8.2 Preparation of Hoagland/Arnon Nutrient Solutions for Nutrient Deficiency Symptoms Development[a]

Stock solution (g/L)	Complete	–N	–P	–K	–Ca	–Mg	–S	–Fe
Major element								
1 M Ca(NO$_3$)$_2$·4H$_2$O (236)	5	—	4	5	—	4	4	5
1 M KNO$_3$ (101)	5	—	6	—	5	6	6	5
1 M KH$_2$PO$_4$ (136)	1	—	—	—	1	1	1	1
1 M MgSO$_4$·7H$_2$O (246)	2	2	2	2	2	—	—	2
Micronutrients[b]	**1**	**1**	**1**	**1**	**1**	**1**	**1**	**1**
50 mM FeDPTA (18.4)[c]	1	1	1	1	1	1	1	—
0.05 M K$_2$SO$_4$ (8.7)	—	5	—	—	—	—	—	—
0.01 M CaSO$_4$·2H$_2$O (1.72)	—	200	—	—	—	—	—	—
0.05 M Ca(H$_2$PO$_4$)$_2$·H$_2$O (12.6)	10	—	10	—	—	—	—	—
1 M Mg(NO$_3$)$_2$·6H$_2$O (256)	—	—	—	—	—	—	2	—

Source: Hoagland, D. R. and Arnon, D. I., 1950, The water culture method for growing plants without soil, circular 347, California Agricultural Experiment Station, University of California, Berkeley.

[a] Liter stock solution per liter nutrient solution.
[b] Contains the following: 2.86 mL/L H$_3$BO$_3$; 1.18 mL/L MnCl$_2$·4H$_2$O; 0.22 mL/L ZnSO$_4$·7H$_2$O; 0.08 mL/L CuSO$_4$·5H$_2$O; 0.02 mL/L H$_2$MoO$_4$·H$_2$O (85% molybdic acid).
[c] Ferric–sodium salt of diethylenediaminetetraacetic acid (DTPA). Differs from Hoagland recipe, which uses iron tartrate.

may be a sufficient quantity of a micronutrient in the plant itself (acquired from the seed) to satisfy the requirement until the plant reaches full maturity. However, it may be worth a try if you like a challenge.

Upon reaching step 10 in the procedure list, the composition of the nutrient solution is altered to free it of one of the essential elements, as shown in Table 8.2.

For B-, Cu-, Mn-, Mo-, and Zn-deficient solutions, substitute micronutrient stock solutions for one of the five salts in the regular micronutrient stock solution. For chlorine-deficient micronutrient solution, substitute 1.55 g MnSO$_4$·H$_2$O for 1.18 g MnCl$_2$·2H$_2$O.

Procedure

1. Remove the bottle from the plastic box and leach the perlite in the bottle with pure water until there is a free flow of water from the hole in the bottom of the bottle. This leaching procedure will free the perlite from any accumulated nutrient solution in the bottle.

2. When the flow of water from the bottom of the bottle ceases, place the bottle into the refrigerator box containing one of the treatment nutrient solutions derived from Table 8.2 (a nutrient solution free from one of the major elements). Be sure to keep at least one bottle on the "full" treatment so that a visual comparison can be made between a plant receiving all of the essential elements versus those missing one of the essential elements.
3. Place the bottle back into its refrigerator box and then back into full sunlight.
4. Maintain the nutrient solution level in the box by adding nutrient solution periodically (usually daily) as required and record the milliliters of solution required to bring back the nutrient solution level to the scribe mark on the side of the box.
5. Depending on the light conditions and rate of growth, significant changes in plant appearance should become evident in about 10 days to 2 weeks.
6. The first evidence of deficiency will be slowed growth.

Photographic record

It would be useful to have a photographic record of the plants at critical stages of development. A daily record can be expensive if a film camera is used; therefore, a digital camera would be the best choice. In order to obtain a meaningful visual record, plant and camera placement is critical. A simple backdrop, called a studio box, can be constructed from a large cardboard box and a piece of blue burlap cloth.

Cut the cardboard box on the diagonal and line the inside of the box with blue burlap cloth on the bottom and up the inside of the box, cutting just to fit the inside box opening. Take one of the box–bottle containers and place it in the center of the bottom of the cardboard box. Draw a square around the box–bottle container, which will designate where the box–bottle should be placed each time a photograph is to be taken. Be sure also to place a mark on the side of the box–bottle container so that it is always oriented in the same way when the photograph is taken.

On the back side of the box, cut small holes just on the inside edge of the back side of the box at 2 in. intervals. Using thick white string, pull a length of string through the holes, creating a series of white lines of string at 2 in. intervals up the inside back of the studio box; this will provide a 2 in. measuring backdrop. A picture of such a constructed box and a box–bottle container in place is shown in Figure 8.3.

For those who have access to a video camcorder, a similar photographic record can be made using daily short exposures of the plants placed in the studio box; be certain that each day's exposure is exactly positioned (video camera and plant). The short daily exposures can then

Figure 8.3 Box and bottle set in the studio box background for photographing.

be edited to give a time-lapse record of the plant as the deficiency symptoms develop.

Plant growth record

A daily record should be kept, observing water use and plant growth. Height measurements may be of little value since, for example, the development of lateral branches is the primary indicator of plant growth for green bean, whereas plant height would be the proper measurement for corn. As the deficiency develops, changes in plant growth will also be influenced by environmental conditions, such as light and temperature. The interaction between these environmental factors and the developing deficiency symptoms can make for an interesting study.

References

Hydroponic reference books

Barry, C. 1996. Nutrients: *The handbook of hydroponics nutrient solutions*. New South Wales, Australia: Casper Publications Pty Ltd.

Bentley, M. 1974. *Hydroponics plus*. Sioux Falls, SD: O'Conners Printers.

Bradley, P and Tabares, C. H. M. 2000. *Spreading simplified hydroponics: Home hydroponic gardens*. Corvallis, OR: Global Hydroponic Network.

Cooper, A. 1995. *The ABC's of NFT: nutrient film technique*. Narrabeen, Australia: Casper Publications.

Day, D. 1991. *Growing in perlite, grower digest 12*. London: Grower Books.

Eastwood, T. 1947. *Soilless growth of plants*, 2nd edition. New York: Reinhold Publishing.

Harris, D. 1977. *Hydroponic gardening without soil*. Cape Town, South Africa: Pumell & Sons.

Hewitt, E. J. 1966. Sand and water culture methods in study of plant nutrition. Technical Communications No. 22 (revised). Commonwealth Agricultural Bureau of Horticulture and Plantation Crops: East Malling, Maidstone, Kent, England.

Jones, Jr., J. B. 1997. *Hydroponics: A guide for the hydroponic and soilless culture grower*. Boca Raton, Fl: CRC Press.

Jones, Jr., J. B. 2012b. *Hydroponic handbook: How hydroponic growing systems work*. Anderson, SC: GroSystems, Inc.

Muckle, E. M. 1993. *Hydroponic nutrients: Easy way to make your own*. Princeton, British Columbia, Canada: Growers Press, Inc.

Resh, H. M. 1998. *Hydroponics: Questions and answers for successful growing*. Santa Barbara, CA: Woodbridge Press.

Resh, H. M. 2002a. *Hobby hydroponics*, 2nd edition, Boca Raton, FL: CRC Press

Resh, H. M. 2002b. *Hydroponic food production*, 6th edition, Boca Raton, FL: CRC Press.

Roberto, K. 2001. *How-to hydroponics*, 3rd ed. Farmingdale, NY: Futuregarden, Inc.

Savage, A. J. (ed.), 1985. *Hydroponic worldwide: State of the art in soilless crop production*. Honolulu, HI: International Center for Special Studies.

Cited references

Alexander, T. 2003. The 2003 South Pacific Soilless Culture Conference. *Growing Edge* 14 (5): 14–19.

Alexander, T., and Coene, T. 1995–1996. Hobby hydroponics for the cost-conscious grower. *Growing Edge* 7 (2): 28–32, 69.

Ames, M., and Johnson, W. S. 1986. A review of factors affecting plant growth. In *Proceedings 7th Annual Conference on Hydroponics: The Evolving Art, the Evolving Science.* Hydroponic Society of America, Concord, CA.

Angus, J. 1995–1996. Hydroculture: A secret worth sharing. *Growing Edge* 7 (2): 48–55.

Anon. 1978. Is hydroponics the answer? *American Vegetable Grower* 29 (11): 11–14.

Anon. 1991. Best management practices begin with the diagnostic approach. Norcross, GA: Potash & Phosphate Institute.

_____ 1997. Water: What's in it and how to get it out. *Today's Chemistry* 6 (1): 1–19.

Antkowiak, R. I. 1993. More oxygen for your NFT. *Growing Edge* 4 (3): 59–63.

Argo, W. R., and Fisher, P. R. 2003. *Understanding pH management for container-grown crops.* Columbus, OH: Meister Publishing.

Arnon, D. I., and Stout, R. R. 1939. The essentiality of certain elements in minute quantity for plants with special reference to copper. *Plant Physiology* 14: 371–375.

Asher, C. J. 1991. Beneficial elements, functional nutrients, and possible new essential elements. In *Micronutrients in agriculture,* ed. J. J. Mortvedt, 703–723. SSSA book series number 4, Madison, WI: Soil Science Society of America.

Asher, C. J., and Edwards, D. G. 1978a. Critical external concentrations for nutrient deficiency and excess. In *Proceedings 8th International Colloquium, Plant Analysis and Fertilizer Problems,* ed. A. R. Ferguson, B. L. Bialaski, and J. B. Ferguson, 13–28. Information series no. 134. New Zealand Department of Scientific and Industrial Research, Wellington, New Zealand.

_____ 1978b. Relevance of dilute solution culture studies to problems of low fertility tropical soils. In *Mineral nutrition of legumes in tropical and subtropical soils,* eds. C. S. Andrew and E. J. Kamprath, 131–152. Melbourne, Australia: Commonwealth Scientific & Industrial Research Organization.

Barber, S. A., and Bouldin, D. R., eds. 1984. Roots, nutrient and water influx, and plant growth. ASA special publication 136. American Society of Agronomy, Madison, WI.

Barry, W. L. 1985. Nutrient solutions and hydroponics. In *Proceedings of the 6th Annual Conference of Hydroponics, Hydroponic Society of America,* Concord, CA.

Bauerle, W. L. 1984. Bag culture production of greenhouse tomatoes. Ohio State University, OARDC, special publication 108. Wooster, OH.

Belanger, R. R., Bowen, P. A., Ehret, D. L., and Menzies, J. G. 1995. Soluble silicon: Its role in crop and disease management of greenhouse crops. *Plant Diseases* 79 (4): 329–335.

Bennet, W. F., ed., 1993. Nutrient deficiencies and toxicities in crop plants. St. Paul, MN: APS Press, American Phytopathological Society.

Bradley, P. 2003. Spreading the soilless word down south: The Mexican Institute for Simplified Hydroponics. *Growing Edge* 14 (4): 14–19.

Bradley, P. and Tabares, C. H. M. 2000a. Building by design: hydroponics in the developing countries, Part 1. *Growing Edge* 11(5): 40-51.

Bradley, P. and Tabares, C. H. M. 2000b. Building by design: hydroponics in the developing countries, Part 2. *Growing Edge* 11(6): 46-57.

Bradley, P. and Tabares, C. H. M. 2000c. Building by design: hydroponics in the developing countries, Part 3. *Growing Edge* 12(1): 47-57.

Bradley, P. and Tabares, C. H. M. 2000d. *Spreading Simplified Hydroponics: Home hydroponic Gardens.* Global Hydroponics Network: Corvallis, OR.

Brooke, L. L. 1995. A world ahead: The leaders in hydroponic technology. *Growing Edge* 6 (4): 34–39, 70–71.

Brooke, L. L. and Silberstein, O. 1993. Hydroponics in schools: An educational tool. *Growing Edge* 5 (1): 20-23.

Brown, P. M., Welsh, R. M., and Cary, E. E. 1987. Nickel: A micronutrient essential for higher plants. *Plant Physiology* 85: 801–803.

Bruce, R. R., Palls, J. E., Jr., Harper, L. A., and Jones, J. B., Jr. 1980. Water and nutrient element regulation prescription in nonsoil media for greenhouse crop production. *Communications in Soil Science and Plant Analysis* 11 (7): 677–698.

Bugbee, B. 1995. Nutrient management in recirculating hydroponics culture. In *Proceedings of the 16th Annual Conference of Hydroponics,* ed. M. Bates, 15–30. Hydroponic Society of America, San Ramon, CA.

Buyanovsky, G., Gale, J., and Degani, N. 1981. Ultraviolet radiation for the inactivation of microorganisms in hydroponics. *Plant and Soil* 60: 131–136.

Carruthers, S. 1998. The future of hydroponics: A global perspective. *Growing Edge* 19 (2): 5–6.

Carson, E. W., ed. 1974. *The plant root and its environment.* Charlottesville: University Press of Virginia.

Christian, M. 1997. Guerrilla NFT California style. *Growing Edge* 8 (3): 47–53.

_____ 2001. Nutrient solution dosers: Automation for recirculation. *Growing Edge* 12 (1): 68–74.

Clark, R. B. 1982. Nutrient solution: Growth of sorghum and corn in mineral nutrition studies. *Journal of Plant Nutrition* 5 (8): 1003–1030.

_____ 1997. The ins and outs of soilless gardening. *Growing Edge* 8 (4): 34–40.

Collins, W. L., and Jensen, M. N. 1983. Hydroponics, a 1983 technology overview. Environment Research Laboratory, University of Arizona, Tucson.

Cooper, A. 1976. *Nutrient film technique for growing crops.* London: Grower Books.

_____ 1979a. *Commercial applications of NFT.* London: Grower Books.

_____ 1979b. *The ABC of NFT.* London: Grower Books.

_____ 1985. New ABC's of NET. In *Hydroponics worldwide: State of the art in soilless crop production,* ed. A. J. Savage, 180–185. Honolulu, HI: International Center for Special Studies.

_____ 1988. *The ABC of NFT.* London: Grower Books.

_____ 1995. *The ABC of NFT, nutrient film technique.* Narrabeen, Australia: Casper Publications.

Creaser, G. 1996–1997. The hydroponic home(made) unit. *Growing Edge* 8 (2): 43–47, 79.

_____ 1997. Cultivate herbs and savor the returns. *Growing Edge* 8 (3): 68–73.

Cunningham, D. 1997. Everything old is new again: The return of the Gericke system. *Practical Hydroponics & Greenhouses* May/June: 69–72.

DeKorne, J. B. 1992–93, An orchard of lettuce trees: vertical NET system, *Growing Edge* 4(2): 52–55.

Devries, J. 2003. Hydroponics. In *Ball redbook: Greenhouses and equipment,* vol. 1, 17th ed., ed. C. Beytes, 103–114. Batavia, IL: Ball Publishing.

Docauer, J. M. 2004. Growing the Verti-Gro way. *Growing Edge* 15 (5): 50–54.

Edwards, J. 1999. Partners in successful farming: County extension offices help growers market their products. *Growing Edge* 10 (5): 52–61.

Edwards, K. 1985. New NFT breakthroughs and future directions. In *Hydroponics worldwide: State of the art in soilless crop production,* ed. A. J. Savage, 186–199. Honolulu, HI: International Center for Special Studies.

Epstein, E. 1972. *Mineral nutrition of plants: Principles and perspectives.* New York: John Wiley & Sons.

_____ 1994. The anomaly of silicon in plant biology. *Proceedings of National Academy of Sciences* 91:11–17.

Erickson, C. 1990. Hydroponic nutrient management. *Growing Edge* 1 (1): 53–56.

Eskew, D. L., Welsh, R. M., and Norvell, W. A. 1984. Nickel in higher plants: Further evidence for an essential role. *Plant Physiology* 76: 691–693.

Evans, R. D. 1995. Control of microorganism in flowing nutrient solutions. In *Proceedings of the 16th Annual Conference on Hydroponics,* ed. M. Bates, 31–43. Hydroponic Society of America, San Ramon, CA.

Farnhand, D. S., Hasek, R. F., and Paul, J. L. 1985. Water quality, leaflet 2995. Division of Agriculture Science, University of California, Davis.

Faulkner, S. P. 1998a. The modified Steiner solution: A complete nutrient solution. *Growing Edge* 9 (4): 43–49.

_____ 1998b. Slow-release nutrient amendment mixes. *Growing Edge* 10 (1): 87–88.

Garnaud, J-C. 1985. Plastics and plastic products. In *Hydroponics worldwide: State of the art in soilless crop production,* ed. A. J. Savage, 31–35. Honolulu, HI: International Center for Special Studies.

Geraldson, C. M. 1963. Quantity and balance of nutrients required for best yields and quality of tomatoes. *Proceedings of Florida State Horticultural Society* 76: 153–158.

_____ 1982. Tomato and the associated composition of the hydroponic or soil solution. *Journal of Plant Nutrition* 5 (8): 1091–1098.

Gerber, J. M. 1985. Plant growth and nutrient formulas. In *Hydroponics worldwide: State of the art in soilless crop production,* ed. A. J. Savage, 58–59. Honolulu, HI: International Center for Special Studies.

Gerhart, H. A., and Gerhart, R. C. 1992. Commercial vegetable production in a perlite system. *Proceedings of the 13th Annual Conference on Hydroponics,* ed. D Schact, 35–38. Hydroponic Society of America, San Ramon, CA.

Gericke, W. F. 1929. Aquaculture: A means of crop production. *American Journal of Botany* 16: 862.

_____ 1937. Hydroponics crop production in liquid culture media. *Science* 85: 177–178.

_____ 1940. *The complete guide to soilless gardening.* New York: Prentice Hall.

Glass, D. M. 1989. *Plant nutrition: An introduction to current concepts.* Boston: Jones and Bartlett Publishers.

Hankinson, J. 2000. A hydroponic lesson plan. *Growing Edge* 11 (5): 28–37.

Hershey, D. R. 1992. Plant nutrient solution pH changes. *Journal of Biology Education* 26 (2): 107–111.

_____ 1994. Hydroponics for teaching: History and inexpensive equipment. *American Biology Teacher* 56: 111–118.

_____ 1995. *Plant biology science projects.* New York: John Wiley & Sons.

Hoagland, D. R., and Arnon, D. I. 1950. The water culture method for growing plants without soil. Circular 347, California Agricultural Experiment Station, University of California, Berkeley.

Ikeda, H., and Osawa, T. 1981. Nitrate- and ammonium-N absorption by vegetables from nutrient solution containing ammonium nitrate and the resultant change of solution pH. *Japanese Society for Horticultural Science* 50 (2): 225–230.

Ingratta, F. J., Blom, T. J., and Strave, W. A. 1985. Canada: Current research and developments. In *Hydroponics worldwide: State of the art in soilless crop production,* ed. A. J. Savage, 95–102. Honolulu, HI: International Center for Special Studies.

Jacoby, B. 1995. Nutrient uptake by plants. In *Handbook of plant and crop physiology,* ed. M. Pessarakli, 1–22. New York: Marcel Dekker, Inc.

Jensen, M. N. 1981. New developments in hydroponic systems, descriptions, operating characteristics, evaluations, pp. 1–25. In *Proceedings Hydroponic Where Is It Growing?* Hydroponic Society of America: Brentwood, CA.

Jensen, M. N. 1995. Hydroponics of the future. In *Proceedings of the 16th Annual Conference on Hydroponics,* ed. M. Bates, 125–132. Hydroponic Society of America, San Ramon, CA.

———— 1997. Hydroponics. *HortScience* 32 (6): 1018–1021.

Johnson, B. 2002c. Greenhouse nutrient management regulations and treatment options. *Growing Edge* 13 (6): 38–43.

Jones, J. B., Jr. 1980. Construct your own growing machine. *Popular Science* 216 (3): 87.

———— 1998. *Plant nutrition manual.* Boca Raton, FL: CRC Press, Inc.

———— 2001. *Laboratory guide for conducting soil tests and plant analysis.* Boca Raton, FL: CRC Press, Inc.

———— 2012. *Plant nutrition and soil fertility manual,* 2nd ed. Boca Raton, FL: CRC Press.

Jones, Jr., J. B. 2001. *Laboratory guide for conducting soil tests and plant analysis.* CRC Press: Boca Raton, FL.

Jones, Jr., J. B. 2012a. *Plant nutrition and soil fertility manual,* 2nd ed. CRC Press: Boca Raton, FL.

Jones, Jr., J. B. 2012b. *Hydroponic handbook: how hydroponic growing systems work.* GroSystems, Anderson, SC.

Jones, J. B., Jr., and Gibson, P. A. 2001. Excessive solar radiation and tomatoes. *Growing Edge* 12 (6): 60–65.

———— 2002. A growing perspective: Hydroponics, yesterday, today, and tomorrow. *Growing Edge* 13 (3): 50–56.

———— 2003. A new look at nutrient solutions. *Growing Edge* 14 (5): 57–63.

Kabata-Pendias, H. 2000. *Trace elements in soils and plants,* 3rd ed. Boca Raton, FL: CRC Press.

Kinro, G. Y. 2003. Going pro—How to turn a hobby into a career. *Growing Edge* 14 (5): 44–49.

Kratky, B. A. 1996. *Noncirculating hydroponic methods.* Hilo, HI: DPL Hawaii.

Larson, J. E. 1979. Soilless culture at Texas A&M, pp. 46–61, In *Proceedings First Annual Conference on Hydroponics.* The Soilless Alternative Hydroponic Society of America: Brentwood, CA.

Larson, J. E. 1970. Soilless culture at Texas A&M. In *Proceedings of First Annual Conference on Hydroponics: The Soilless Alternative,* 46–61. Hydroponic Society of America, Brentwood, CA.

Lubkeman, D. 1998. Basic atmospheric greenhouse control. *Growing Edge* 0(6): 57–65.

Lubkeman, D. 1999. Computer controls in the greenhouse. *Growing Edge* 10(3): 63–70.

Markert, B. 1994. Trace element content of "reference plant." In *Biochemistry of trace elements*, eds. D. C. Adriano, Z. S. Chen, and S. S. Yang. Northwood, NY: Science and Technology Letters.

Mengel, K., and Kirkby, E. A. 1987. *Principles of plant nutrition*, 4th ed. Worblaufen-Bern, Switzerland: International Potash Institute.

Mertz, W. 1981. The essential trace elements. *Science* 213: 1332–1338.

Mills, H. A., and Jones, J. B., Jr. 1996. *Plant nutrition handbook II*. Athens, GA: MicroMacro Publishing.

Morgan, L. 1998. The pH factor in hydroponics. Coconut fiber. *Growing Edge* 9 (4): 25–33.

_____ 1999a. Coconut fiber: The environmentally friendly medium. *Growing Edge* 10 (5): 24–30.

_____ 1999b. Introduction to hydroponic gullies and channels. *Growing Edge* 10 (6): 67–75.

_____ 2000. Beneficial elements for hydroponics: A new look at plant nutrition. *Growing Edge* 11 (3): 40–51.

_____ 2002a. Hydroponic classroom experiments. *Growing Edge* 13 (6): 56–70.

_____ 2002b. Hydroponic Q&A. *Growing Edge* 14 (1): 11.

_____ 2002c. Raft system specifics. *Growing Edge* 14 (2): 46–60.

Morgan, L. 2003a. Carbon dioxide enrichment. *Growing Edge* 14 (5): 64–74.

Morgan, L. 2003b. Hydroponic substrates. *Growing Edge* 15 (2): 54–66.

Musgrave, C. E. 2001. Creating your own nutrient solution. *Growing Edge* 13 (1): 60–66.

Naegely, S. K. 1997. Greenhouse vegetables business is booming. *Greenhouse Grower* 15: 14–18.

Nederhoff, E. 2001. Commercial greenhouse growing in the Netherlands. *Growing Edge* 12 (4): 31–43.

Nichols, M. 2002. Aeroponics: Production systems and research tools. *Growing Edge* 13 (5): 30–35.

Pais, I. 1992. Criteria of essentiality, beneficiality, and toxicity of chemical elements. *Acta Alimentaria* 21 (2): 145–152.

Pallas, J. E., Jr., and Jones, J. B., Jr. 1978. Platinum uptake by horticultural crops. *Plant and Soil* 50: 207–212.

Peckenpaugh, D. J. 2002a. Hobbyist hydroponics: Some general resources for every grower. *Growing Edge* 13 (4): 31–39.

_____ 2003a. Homemade recirculating drip hydroponics: Part 1: System construction. *Growing Edge* 14 (6): 75–78.

_____ 2003b. Homemade recirculating drip hydroponics: Part 2: System operation and maintenance. *Growing Edge* 15 (2): 74–77.

Pokorny, F. A. 1979. Pine bark container media—An overview. *Combined Proceedings of International Plant Propagators Society* 29: 484–495.

Rengel, Z. 2002. Chelator EDTA in nutrient solution decreases growth of wheat. *J. plant nutrition* 25 (8): 1709–1725.

Rodriguez de Cianzio, S. R. 1991. Recent advances in breeding for improving iron utilization by plants. In *Iron nutrition and interactions in plants*, eds. Y. Chen and Y. Hadar, 83–88. Dordrecht, The Netherlands: Kluwer Academic Publishers.

Rorabaugh, P. A. 1995. A brief and practical trek through the world of hydroponics. In *Proceedings of the 16th Annual Conference on Hydroponics*, ed. M. Bates, 7–14. Hydroponic Society of America, San Ramon, CA.

Russell, E. J. 1950. *Soil conditions and plant growth.* London: Longmans, Green and Company.

Savage, A. J., ed., 1985. *Hydroponics worldwide: State of the art in soilless crop production.* Honolulu, HI: International Center for Special Studies.

Schippers, P. A. 1979. The nutrient flow technique. V. C. Mimeo 212. Department of Vegetable Crops, Cornell University, Ithaca, NY.

Schmitz, J. 2003. Couple spreads hydroponic gospel in the Northwest. *Fruit Growers News* 43 (4): 40.

Schmitz, J. 2004. Couple spreads hydroponics gospel in the Northwest. *The Fruit Growers News* 43 (1): 40.

Schneider, R. 1998. Winter tomatoes at the office. *Growing Edge* 9 (6): 17–19.

_____ 2000. The hydroponic adventure continues—Creative chaos and catastrophe. *Growing Edge* 11 (3): 17–23.

_____ 2002. Summer salad system success. *Growing Edge* 13 (3): 44–49.

_____ 2003, Summer system VI: The birth of Star Hydro. *Growing Edge* 14 (3): 75–80.

_____ 2004. Summer's harvest interrupted. *Growing Edge* 15 (3): 69–76.

Schneider, R., and Ericson, L. 2001. Backyard buckeye hydro. *Growing Edge* 12 (4): 75–79.

Schoenstein, G. P. 2001. Hope through hydroponics. *Growing Edge* 13 (2): 69–79.

Schon, M. 1992. Tailoring nutrient solution to meet the demands of your plants. In *Proceedings of the 13th Annual Conference on Hydroponics,* ed. D. Schact, 1–7. Hydroponic Society of America, San Ramon, CA.

Schwarz, M. 2003. Industry perspectives: Israeli pioneers of hydroculture. *Growing Edge* 15 (1): 52–58.

Silberstein, O. 1995. Hydroponics: Making waves in the classroom. *Growing Edge* 6 (4): 16.

Silberstein, O., and Spoelstra-Pepper, C. 1999. Hydroponic workshops for teachers: A traveling road show. *Growing Edge* 11 (2): 10–11.

Simon, D. 2004. Hydro machine goes round and round. *Growing Edge* 15 (5): 10–11.

Smith, B. 2001a. Designing and building your own home hydroponic system: Part one: Let's kick around some possibilities! *Growing Edge* 12 (3): 78–80.

_____ 2001b. Designing and building your own home hydroponic system: Part two: Let's keep it simple…at the start! *Growing Edge* 12 (4): 80–83.

Smith, B. 2001c. Designing and building your own home hydroponic system. Part two: Let's keep It simple. *Growing Edge* 12 (5): 74–79.

Smith, B., 2001d. Designing and building your own home hydroponic system. Part three: Let's get growing. *Growing Edge* 12 (6): 78–83.

_____ 2004. A short history of NFT gully design. *Growing Edge* 15 (3): 79–82.

Smith, D. 1987. *Rock wool in horticulture.* London: Grower Books.

Smith, R. 1999. The growing world of hydroponics. *Growing Edge* 11 (1): 14–16.

Sonneveld, C. 1989. Rockwool as a substance in protected cultivation. *Chronica Horticulturae* 29 (3): 33–38.

Spillane, M. 2001. Fresh greens from Quebec. *Growing Edge* 12 (6): 52–59.

_____ 2002. Reusing rockwool: Economical and environmental solutions for commercial growers. *Growing Edge* 13 (6): 30–37.

Steiner, A. A. 1961. A universal method for preparing nutrient solutions of certain desired composition. *Plant and Soil* 15: 134–154.

_____ 1980. The selective capacity of plants for ions and its importance for the composition of the nutrient solution. In *Symposium on Research on Recirculating Water Culture*, eds. R. G. Hurd, P. Adams, D. M. Massey, and D. Price, 37–97. *Acta Horticulture* no. 98. Netherlands: The Hague.

_____ 1984. The universal nutrient solution. In *Proceedings of Sixth International Congress of Soilless Culture*, 633–650. The Hague, The Netherlands.

_____ 1985. The history of mineral plant nutrition till about 1860 as source of the origin of soilless culture methods. *Soilless Culture* 1 (1): 7–24.

Takahashi, E., Ma, J. F., and Miyake, Y. 1990. The possibility of silicon as an essential element for higher plants. In *Comments on agricultural and food chemistry*, 99–122. London: Gordon and Breach Scientific Publications.

Tindall, J. A., Mills, H. A., and Radcliffe, D. E. 1990. The effect of root zone temperature on nutrient solution uptake of tomato. *Journal of Plant Nutrition* 13: 939–956.

Trelease, S. F., and Trelease, H. M. 1935. Physiologically balanced culture solutions with stable hydrogen ion concentration. *Science* 78: 438–439.

Van Patten, G. F. 1992. Hydroponics for the rest of us. *Growing Edge* 3 (3): 24–33, 48–51.

Waters, W. F., Geraldson, C. M., and Woltz, S. S. 1972. *The interpretation of soluble salt tests and soil analysis by different procedures*. AREC. Mimeo. Report GC-1072. Bradenton: FL.

Whipker, B. E., Dole, J. M., Cavins, T. J., and Gobson, J. L. 2003. Water quality. In *Ball redbook: Crop production*, vol. 2, 17th ed., ed. D. Hamrick, 9–18. Batavia, IL: Ball Publishing.

Wignarajah, K. 1995. Mineral nutrition in plants. In *Handbook of plant and crop physiology*, ed. M. Pessarakli, 193–222. New York: Marcel Dekker.

Wilcox, G. E. 1980. High hopes for hydroponics. *American Vegetable Grower* 28: 11–14.

_____ 1983. Hydroponic systems around the world, their characteristics and why they are used. In *Proceedings of Fourth Annual Conference, Theme: Hydroponics How Does It Work?*, 1–14. Hydroponic Society of America, Concord, CA.

Wilson, G. 2002a. Soilless systems in the sky: To boldly grow where none have grown before. *Growing Edge* 13 (3): 24–29.

_____ 2002b. Expanding into aeroponics. *Growing Edge* 13 (5): 36–39.

Wittwer, S. H. 1993. Worldwide use of plastics in horticultural crops. *HortTechnology* 3: 6–19.

Yuste, M.-P., and Gostincar, J., eds. 1999. *Handbook of agriculture*. New York: Marcel Dekker.

Appendix A: Measurement factors

Common prefixes

Factor	Prefix	Symbol
1,000,000	Mega	M
1,000	Kilo	k
1/100	Centi	c
1/1000	Milli	m
1/1,000,000	Micro	μ

Metric conversion factors (approximate)

When you know		Multiply by	To find	Symbol
Length	Inches	2.54	Centimeters	cm
	Feet	30	Centimeters	cm
	Yards	0.9	Meters	m
	Miles	1.6	Kilometers	km
Area	Square inches	6.5	Square centimeters	cm²
	Square feet	0.09	Square meters	m²
	Square yards	0.8	Square meters	m²
	Square miles	2.6	Square kilometers	km²
	Acres	0.4	Hectares	ha
Weight	Ounces	28	Grams	g

(Continued)

When you know		Multiply by	To find	Symbol
	Pounds	0.45	Kilograms	kg
	Short tons (2000 pounds)	0.9	Metric tons	t
Volume	Teaspoons	5	Milliliters	mL
	Tablespoons	15	Milliliters	mL
	Cubic inches	16	Milliliters	mL
	Fluid ounces	30	Milliliters	mL
	Cups	0.24	Liters	L
	Pints	0.47	Liters	L
	Quarts	0.95	Liters	L
	Gallons	3.8	Liters	L
	Cubic feet	0.03	Cubic meters	m^3
	Cubic yards	0.76	Cubic meters	m^3
Pressure	Inches of mercury	3.4	Kilopascals	kPa
	Pounds/square inch	6.9	Kilopascals	kPa
Temperature (exact)	Degrees Fahrenheit	5/9 (after subtracting 32)	Degrees Celsius	°C

Useful information and conversion factors

Name	Symbol	Approximate size or equivalent
	Length	
Meter	m	39.5 inches
Kilometer	km	0.6 mile
Centimeter	cm	Width of a paper clip
Millimeter	mm	Thickness of a paper clip
	Area	
Hectare	ha	2.5 acres
	Weight	
Gram	g	Weight of a paper clip
Kilogram	kg	2.2 pounds
Metric ton	t	Long ton (2,240 pounds)
	Volume	
Liter	L	1 quart and 2 ounces
Milliliter	mL	1/5 teaspoon

(Continued)

Pressure

Kilopascal	kPa	Atmospheric pressure is about 100 kPa

Temperature

Celsius	°C	5/9 after subtracting 32 from °F
Freezing	0°C	32°F
Boiling	100°C	212°F
Body temp.	37°C	98.6°F
Room temp.	20°C to 25°C	68°F to 77°F

Electricity

Kilowatt	kW
Kilowatt-hour	kWh
Megawatt	MW

Miscellaneous

Hertz	Hz	One cycle per second

Yield or rat

Ounces per acre (oz/acre) × 0.07 = Kilograms per hectare (kg/ha)
Tons per acre (ton/acre) × 2240 = Kilograms per hectare (kg/ha)
Tons per acre (ton/acre) × 2.24 = Metric tons per hectare (kg/ha)
Pounds per acre (lb/acre) × 1.12 = Kilograms per hectare (kg/ha)
Pounds per cubic foot (lb/ft^3) × 16.23 = Kilograms per cubic meter (kg/m^3)
Pounds per gallon (lb/gal) × 0.12 = Kilograms per liter (kg/L)
Pounds per ton (lb/ton) × 0.50 = Kilograms per metric ton (kg/MT)
Gallons per acre (gal/acre) × 9.42 = Liters per hectare (L/ha)
Gallons per ton (gal/ton) × 4.16 = Liters per metric ton (L/MT)
Pounds per 100 square feet (lb/100 ft^2) × 2 = Pounds/100 gallons water
  (assumes that 100 gallons will saturate 200 square feet of soil)
Pounds per acre (lb/acre)/43.56 = Pounds per 1000 square feet (lb/1000 ft^2)

Volumes and liquids

Teaspoon (tsp) = 1/2 tablespoon = 1/16 ounce (oz)
Tablespoon (tbs) = 3 teaspoons = 1/2 ounce
Fluid ounces (fl oz) = 2 tablespoons = 6 teaspoons
Pint/100 gallons (gal) = 1 teaspoon per gallon
Quart per 100 gallons = 2 tablespoons per gallon
3 teaspoons = 1 tablespoon = 14.8 milliliters (mL)
2 tablespoons = 1 fluid ounce = 29.6 milliliters
8 fluid ounces = 16 tablespoons = 1 cup = 236.6 milliliters
2 cups = 32 tablespoons = 1 pint = 473.1 milliliters

(Continued)

2 pints = 64 tablespoons = 1 quart (qt) = 946.2 milliliters
1 liter (L) = 1,000 milliliters = 1,000 cubic centimeters (cc) = 0.264
Gallons = 33.81 ounces
4 quarts = 256 tablespoons = 1 gallon = 3,785 milliliters
1 gallon = 128 ounces = 3.785 liters

Elemental conversions

$P_2O_5 \times 0.437$ = Elemental P   Elemental P $\times 2.29 = P_2O_5$
$K_2O \times 0.826$ = Elemental K   Elemental K $\times 1.21 = K_2O$
$CaO \times 1.71$ = Elemental Ca   Elemental Ca $\times 1.40 = CaO$
$MgO \times 0.60$ = Elemental Mg   Elemental Mg $\times 1.67 = MgO$
$CaCO_3 \times 0.40$ = Elemental Ca

Weight/mass

Ounce (oz) = 28.35 grams (g)
16 ounces = 1 pound (lb) = 453.6 grams
Kilogram (kg) = 1,000 grams = 2.205 pounds
Gallon (gal) water = 8.34 pounds = 3.8 kilograms
1 cubic foot (ft^3) of water = 62.4 pounds = 28.3 kilograms
1 kilogram of water = 33.81 ounces
Ton (t) = 2,000 pounds = 907 kilograms
1 metric ton (MT) = 1,000 kilograms = 2,205 pounds

Volume equivalents

Gallon in 100 gallons = 1 1/4 ounces in 1 gallon
Quart in 100 gallons = 5/16 ounce in 1 gallon
1 pint in 100 gallons = 1/16 ounce in 1 gallon
8 ounces in 100 gallons = 1/2 teaspoon in 1 gallon
4 ounces in 100 gallons = 1/4 teaspoon in 1 gallon

Temperature			
°C	°F	°C	°F
5	40	120	248
10	50	125	257
19.4	67	180	356
20	68	200	392
21	70	330	626
23	73	350	662
25	77	370	698
27	80	400	752
32	90	450	842

(Continued)

38	100	500	932
40	105	550	1022
50	122	600	1122
80	176	900	1652
100	212	1350	2462
110	230		

$$°F = (°C + 17.78) \times 1.8$$
$$°C = (°F - 32) \times 0.556$$

Comparison of commonly used concentration units for major elements and micronutrients in dry matter of plant tissue

Elements	Concentration units[a]			
	%	g/kg	cmol(p+)/kg	cmol/kg
Major elements				
Phosphorus (P)	0.32	5.2	—	10
Potassium (K)	1.95	19.5	50	50
Calcium (Ca)	2.00	20.0	25	50
Magnesium (Mg)	0.48	4.8	10	20
Sulfur (S)	0.32	3.2	—	10
Micronutrients	ppm	mg/kg	mmol/kg	
Boron (B)	20	20	1.85	
Chlorine (Cl)	110	100	2.82	
Iron (Fe)	111	111	1.98	
Manganese (Mn)	55	55	1.00	
Molybdenum (Mo)	1	1	0.01	
Zinc (Zn)	33	33	0.50	

[a] Concentration levels were selected for illustrative purposes only.

To convert molar units to parts per million

Multiply the millimoles per liter (mmol/L) or micromoles per liter (μmol/L) by the atomic weight of the element = parts per million (ppm).

	Symbol	Atomic weight	mmol/L[a]	ppm
Macroelement (millimoles/liter)				
Nitrogen	N	14	17.25	242
Potassium	K	39	11.00	429
Phosphorus	P	31	2.25	70
Calcium	Ca	40	5.5	220
Magnesium	Mg	24	1.50	36
Sulfur	S	32	2.27	88
Macroelement (micromoles/liter)				
Boron	B	11	31.25	0.34
Copper	Cu	64	0.625	0.04
Iron	Fe	56	1.25	7.00
Manganese	Mn	55	43.75	2.40
Molybdenum	Mo	96	0.625	0.06
Zinc	Zn	65	1.875	0.12

[a] Chosen for illustrative purposes.

To convert pounds per acre to milliequivalents per 100 g

Element	Multiply by
Calcium (Ca)	400
Magnesium (Mg)	780
Potassium (K)	240
Sodium (Na)	460

Conversion values useful for completing nutrient solution calculations

1.0 pound (lb)	= 454 grams (g)
2.2 pounds	= 1 kilogram (kg)
1.0 gram	= 1000 milligrams (mg)
1.0 gallon (gal)	= 3.78 liters (L)
1.0 liter	= 1,000 milliliters (mL)
1.0 milligram/liter	= 1 part per million (ppm)
1.0 pound	= 16 ounces (oz)

(Continued)

1.0 gallon water	= 8.3 pounds
1.0 quart (qt)	= 0.95 liters
1.0 gallon	= 128 ounces
1.0 gallon	= 3,780 milliliters

Units of measurement for electrical conductivity and related terms

- Decisiemens per meter (dS/m)
- Millisiemens per centimeter (mS/cm)
- Microsiemens per centimeter (μS/cm)
- 1 dS/m = 1 mS/cm = 1000 μS/cm = 1 mmho/cm
- 1 μS/cm = 0.001 dS/m
- EC (in dS/m) × 640 = TDS (in mg/L [ppm])

Notes: approximate measurement, depends on type of salt; cf. (conductivity factor) of 10 = 1 dS/m.

Appendix B: Essential element summarization tables

In this appendix, the characteristics of the essential elements are presented in outline form for easy reference. The objective is to provide the most useful information about each essential element in one easy-to-follow format. The information and data given are primarily in reference to the hydroponic/soilless growing methods for those crops thus commonly grown; therefore, the information given may not be useful for application with other growing methods or crops. The critical and excessive levels and the sufficiency ranges for the essential elements have been selected as probable levels and should not be considered specific. These levels are what would be found in recently mature leaves, unless otherwise specified.

Nitrogen (N)

Atomic number: 7   **Atomic weight:** 14.00
Discoverer of essentiality and year: DeSaussure, 1804
Designated element: major element
Function: used by plants to synthesize amino acids and form proteins, nucleic acids, alkaloids, chlorophyll, purine bases, and enzymes
Mobility: mobile
Forms utilized by plants: nitrate (NO_3^-) anion and ammonium (NH_4^+) cation

Common reagent sources for making nutrient solutions

Reagent	Formula	% N
Ammonium dihydrogen phosphate	$NH_4H_2PO_4$	11 (21% P)
Ammonium hydroxide	NH_4OH	20–25
Ammonium nitrate	NH_4NO_3	32 (16% NH_4 and 16% NO_3)
Ammonium sulfate	$NH_4(SO_4)_2$	21 (24% S)
Diammonium hydrogen phosphate	$(NH_4)_2HPO_4$	18 (21% P)
Calcium nitrate	$Ca(NO_3)_2 4H_2O$	15 (19% Ca)
Potassium nitrate	KNO_3	13 (36% K)

Concentration in nutrient solutions: 100 to 200 mg/L (ppm); in a NO_3-based nutrient solution having 5% to 10% of the N as NH_4 will increase the uptake of N

Typical deficiency symptoms: very slow growing, weak, and stunted plants; leaves light green to yellow in color, beginning with the older leaves; plants mature early, and dry weight and fruit yield are reduced

Plant symptoms of excess: plants dark green in color with succulent foliage; easily susceptible to environmental stress and disease and insect invasion; poor fruit yield of low quality

Critical plant levels: 3.00% total N (will vary with plant type and stage of growth); 1000 mg/kg (ppm) NO_3–N in leaf petiole

Excessive plant level: >5.00% total N (will vary with plant type and stage of growth); >2000 mg/kg (ppm) NO_3–N in leaf petiole

Ammonium toxicity: when NH_4 is the major source of N, toxicity can occur, seen as cupping of plant leaves, breakdown of vascular tissue at the base of the plant, lesions on stems and leaves, and increased occurrence of blossom-end rot (BER) on fruit

Phosphorus (P)

Atomic number: 15   Atomic weight: 30.973

Discoverer of essentiality and year: Ville, 1860

Designated element: major element

Function: component of certain enzymes and proteins involved in energy transfer reactions and component of RNA and DNA

Mobility: mobile

Forms utilized by plants: mono- and dihydrogen phosphates ($H_2PO_4^-$ and HPO_4^{2-}, respectively) anions, depending on pH

Common reagents for making nutrient solutions

Reagent	Formula	% P
Ammonium dihydrogen phosphate	$NH_4H_2PO_4$	21 (11% N)
Diammonium hydrogen phosphate	$(NH_4)_2HPO_4$	21 (81% N)
Dipotassium hydrogen phosphate	K_2HPO_4	18 (22% N)
Phosphoric acid	H_3PO_4	34
Potassium dihydrogen phosphate	KH_2PO_4	32 (30% K)

Concentration in nutrient solutions: 30 to 50 mg/L (ppm) (The author recommends that the P content in a nutrient solution be between 10 and 20 mg/L or ppm.)

Typical deficiency symptoms: slow and reduced growth, with developing purple pigmentation of older leaves; foliage very dark green in color

Symptoms of excess: plant growth will be slow, with some visual symptoms possibly related to a micronutrient deficiency, such as Zn

Critical plant level: 0.25% total; 500 mg/kg (ppm) extractable P in leaf petiole

Excessive plant level: >1.00% total; >3000 mg/kg (ppm) extractable P in leaf petiole

Potassium (K)

Atomic number: 19   Atomic weight: 39.098
Discoverer of essentiality and year: von Sachs, Knop, 1860
Designated element: major element
Function: maintains the ionic balance and water status in plants, involved in the opening and closing of stomata, and associated with carbohydrate chemistry
Mobility: mobile
Form utilized by plants: potassium (K⁺) cation

Common reagents for making nutrient solutions

Reagent	Formula	% K
Dipotassium hydrogen phosphate	K_2HPO_4	22 (18% P)
Potassium chloride	KCl	50 (47% Cl)
Potassium dihydrogen phosphate	KH_2PO_4	30 (32% P)
Potassium nitrate	KNO_3	36 (13% N)
Potassium sulfate	K_2SO_4	42 (17% S)

Concentration in nutrient solutions: 100 to 200 mg/L (ppm)

Typical deficiency symptoms: initially slowed growth with marginal death of older leaves giving a burned or scorched appearance; fruit yield and quality reduced; fruit postharvest quality reduced

Symptoms of excess: plants will develop either Mg or Ca deficiency symptoms; plants can take up K easily and the amount found in the plant may exceed the biological need, called "luxury consumption"

Critical plant level: 2.00%

Excessive plant level: >6.00%, which will be less depending on plant type and stage of growth; however, there are plants that have high K requirements greater than 6.00%

Calcium (Ca)

Atomic number: 20   **Atomic weight:** 40.07

Discoverer of essentiality and year: von Sachs, Knop, 1860

Designated element: major element

Functions: major constituent of cell walls and for maintaining cell wall integrity and membrane permeability; enhances pollen germination and growth; activates a number of enzymes for cell mitosis, division, and elongation; may detoxify the presence of heavy metals in tissue

Mobility: immobile

Form utilized by plants: calcium (Ca^{2+}) cation

Common reagents for making nutrient solutions

Reagent	Formula	% Ca, dry weight
Calcium chloride	$CaCl_2$	36 (64% Cl)
Calcium nitrate	$Ca(NO_3)2 \cdot 4H_2O$	19 (15% N)
Calcium sulfate	$CaSO_4 \cdot 2H_2O$	23 (19% S)

Concentration in nutrient solutions: 200 to 300 mg/L (ppm)

Typical deficiency symptoms: leaf shape and appearance will change, with the leaf margins and tips turning brown or black; edges of leaves may look torn; vascular breakdown at the base of the plant; for fruit crops, occurrence of blossom-end rot (BER)

Symptoms of excess: may induce possible Mg or K deficiency

Critical plant level: 1.00% (will vary with plant type and stage of growth)

Excessive plant level: >5.00% (will vary with level of K and/or Mg)

Magnesium (Mg)

Atomic number: 12   **Atomic weight:** 24.30

Discoverer of essentiality and year: von Sachs, Knop, 1860

Designated element: major element

Functions: major constituent of the chlorophyll molecule; enzyme activator for a number of energy transfer reactions

Mobility: moderately mobile

Form utilized by plants: magnesium (Mg^{2+}) cation

Common reagent for making nutrient solutions

Reagent	Formula	% Mg
Magnesium sulfate	$MgSO_4 \cdot 7H_2O$	10 (23% S)

Concentration in nutrient solutions: 30 to 50 mg/L (ppm)
Typical deficiency symptoms: interveinal chlorosis on older leaves; possible development of blossom-end rot in fruit
Symptoms of excess: results in cation imbalance among Ca and K; slowed growth with the possible development of either Ca or K deficiency symptoms
Critical plant level: 0.25%
Excessive plant level: >1.50% (will vary with level of K and/or Ca)

Sulfur (S)

Atomic number: 16   **Atomic weight:** 32.06
Discoverer of essentiality and year: von Sachs, Knop, 1865
Designated element: major element
Functions: constituent of two amino acids: cystine and thiamine; component of compounds that give unique odor and taste to some types of plants
Mobility: moderately mobile
Form utilized by plants: sulfate (SO_4^{2-}) anion

Common reagents for making nutrient solutions

Reagent	Formula	% S
Ammonium sulfate	$(NH_4)_2SO_4$	24 (21% N)
Calcium sulfate	$CaSO_4 \cdot 2H_2O$	23 (26% Ca)
Magnesium sulfate	$MgSO_4 \cdot 7H_2O$	23 (10% Mg)
Potassium sulfate	K_2SO_4	17 (42% K)

Concentration in nutrient solutions: 70 to 150 mg/L (ppm)
Typical deficiency symptoms: general loss of green color of the entire plant; slowed growth
Symptoms of excess: not well defined
Critical plant level: 0.30%

Boron (B)

Atomic number: 5   **Atomic weight:** 10.81
Discoverer of essentiality and year: Sommer and Lipman, 1926
Designated element: micronutrient
Functions: associated with carbohydrate chemistry, pollen germination, and cellular activities (division, differentiation, maturation, respiration, and growth); important in the synthesis of one of the bases for RNA formation
Mobility: immobile
Forms utilized by plants: borate (BO_3^{3-}) anion as well as the molecule H_3BO_3

Common reagents for making nutrient solutions

Reagent	Formula	% B
Boric acid	H_3BO_3	16
Solubor	$Na_2B_4O_7 \cdot 4H_2O + Na_2B_{10}O_{16} \cdot 10H_2O$	20
Borax	$Na_2B_4O_7 \cdot 10H_2O$	11

Concentration in nutrient solutions: 0.3 mg/L (ppm)

Typical deficiency symptoms: slowed and stunted new growth, with possible death of the growing point and root tips; lack of fruit set and development; plants are brittle and petioles will easily break off the stem

Symptoms of excess: accumulates in the leaf margins, resulting in death of the margins

Critical plant level: 25 mg/kg (ppm)

Toxic plant level: >100 mg/kg (ppm)

Chlorine (Cl)

Atomic number: 17   **Atomic weight:** 35.45

Discoverer of essentiality and year: Stout, 1954

Designated element: micronutrient

Functions: involved in the evolution of oxygen (O_2) in photosystem II; raises cell osmotic pressure and affects stomatal regulation; increases hydration of plant tissue

Mobility: mobile

Form utilized by plants: chloride (Cl⁻) anion

Common reagent for making nutrient solutions

Reagent	Formula	% Cl
Potassium chloride	KCl	47 (50% K)

Concentration in nutrient solutions: 50 to 1000 mg/L (ppm) (depends on reagents used)

Typical deficiency symptoms: chlorosis of the younger leaves; wilting

Symptoms of excess: premature yellowing of leaves; burning of leaf tips and margins; bronzing and abscission of leaves

Critical plant level: 20 mg/kg (ppm)

Excess level: >0.50%

Copper (Cu)

Atomic number: 29   **Atomic weight:** 64.54
Discoverer of essentiality and year: Sommer, 1931
Designated element: micronutrient
Functions: constituent of the chloroplast protein plastocyanin; participates in electron transport system linking photosystems I and II; participates in carbohydrate metabolism and nitrogen (N_2) fixation
Mobility: immobile
Form utilized by plants: cupric (Cu^{2+}) cation

Common reagent for making nutrient solutions

Reagent	Formula	% Cu
Copper sulfate	$CuSO_4 \cdot 5H_2O$	25 (13% S)

Concentration in nutrient solutions: 0.01 to 0.1 mg/L (ppm); highly toxic to roots when in excess of 1.0 mg/L (ppm) in solution
Typical deficiency symptoms: reduced or stunted growth, with a distortion of the young leaves; necrosis of the apical meristem
Symptoms of excess: induced iron deficiency and chlorosis; root growth will cease and root tips will die and turn black
Critical plant level: 5 mg/kg (ppm)
Toxic plant level: >30 mg/kg (ppm)

Iron (Fe)

Atomic number: 26   **Atomic weight:** 55.85
Discoverer of essentiality and year: von Sachs, Knop, 1860
Designated element: micronutrient
Functions: component of many enzyme and electron transport systems; component of protein ferredoxin; required for NO_3 and SO_4 reduction, N_2 assimilation, and energy (NADP) production; associated with chlorophyll formation
Mobility: immobile
Forms utilized by plants: ferrous (Fe^{2+}) and ferric (Fe^{3+}) cations

Common reagents for making nutrient solutions

Reagent	Formula	% Fe
Iron chelate	FeDTPA	6–12
Iron citrate		
Iron tartrate		
Iron lignin sulfonate		6
Ferrous sulfate	$FeSO_4 \cdot 7H_2O$	20 (11% S)
Ferrous ammonium sulfate	$(NH_4)_2SO_4 \cdot FeSO_4\,6H_2O$	14

Concentration in nutrient solutions: 2 to 12 mg/L (ppm)

Typical deficiency symptoms: interveinal chlorosis of younger leaves; as deficiency intensifies, older leaves are affected and younger leaves turn yellow; deficiency can be genetically induced

Symptoms of excess: not known for crops commonly grown hydroponically

Critical plant level: 50 mg/kg (ppm)

Excess plant level: not known

Manganese (Mn)

Atomic number: 25   **Atomic weight:** 54.94

Discoverer of essentiality and year: McHargue, 1922

Designated element: micronutrient

Functions: involved in oxidation–reduction processes in the photosynthetic electron transport system; photosystem II for photolysis; activates IAA oxidases

Mobility: immobile

Form utilized by plants: manganous (Mn^{2+}) cation

Common reagents for making nutrient solutions

Reagent	Formula	% Mn
Manganese sulfate	$MnSO_4 \cdot 4H_2O$	24 (14% S)
Manganese chloride	$MnCl_2 \cdot 4H_2O$	28

Concentration in nutrient solutions: 0.5 to 2.0 mg/L (ppm); high P in the nutrient solution can increase the uptake of Mn

Typical deficiency symptoms: reduced and stunted growth, with interveinal chlorosis on younger leaves

Symptoms of excess: older leaves show brown spots surrounded by chlorotic zone or circle; black spots (called "measles") will appear on stems and petioles

Critical plant level: 25 mg/kg (ppm)

Toxic plant level: >400 mg/kg (ppm)

Molybdenum (Mo)

Atomic number: 42   **Atomic weight:** 95.94
Discoverer of essentiality and year: Sommer and Lipman, 1926
Designated element: micronutrient
Functions: component of two enzyme systems—nitrogenase and nitrate reductase—for the conversion of NO_3 to NH_4
Mobility in plant: immobile
Form utilized by plants: molybdate (MoO_4^-) anion

Common reagent for making nutrient solutions

Reagent	Formula	% Mo
Ammonium molybdate	$(NH_4)_6Mo_7O_{24}\cdot 4H_2O$	8 (1% N)

Concentration in nutrient solutions: 0.05 mg/L (ppm)
Typical deficiency symptoms: resemble N deficiency symptoms, with older and middle leaves becoming chlorotic; leaf margins will roll; growth and flower formation restricted
Symptoms of excess: not known
Critical plant level: not exactly known, but probably 0.10 mg/kg (ppm)
Excess plant level: not known

Zinc (Zn)

Atomic number: 30   **Atomic weight:** 65.39
Discoverer of essentiality and year: Lipman and MacKinnon, 1931
Designated element: micronutrient
Functions: involved in same enzymatic functions as Mn and Mg; specific to the enzyme carbonic anhydrase
Mobility: immobile
Form utilized by plants: zinc (Zn^{2+}) cation

Common reagent for making nutrient solutions

Reagent	Formula	% Zn
Zinc sulfate	$ZnSO_4\cdot 7H_2O$	22 (11% S)

Concentration in nutrient solutions: 0.05 mg/L (ppm), may need to be 0.10 mg/L if a chelated form of Fe is in the formulation; can be highly toxic to roots when in excess of 0.5 mg/L (ppm)

Typical deficiency symptoms: upper new leaves will curl with rosette appearance; chlorosis in the interveinal areas of new leaves produces a banding effect; leaves will die and fall off; flowers will abscise

Symptoms of excess: plants may develop typical Fe deficiency symptoms; chlorosis of young leaves

Critical plant level: 15 mg/kg (ppm)

Toxic plant level: >100 mg/kg (ppm); high Zn can interfere with Fe nutrition

Appendix C: Diagnostic testing

Importance

Success with any growing system is based to a considerable degree on the ability of the grower to evaluate and diagnose the condition of his crop effectively at all times. This is particularly true for any grower, but is absolutely essential for the commercial hydroponic grower. Some growers possess a unique ability to sense when things are not right and take the proper corrective steps before significant crop damage occurs. Most, however, must rely on more obvious and objective measures to assist them in determining how their growing system is functioning and how plants are responding to their management inputs. In the latter case, no substitute for systematic observations and testing exists. As the genetic growth and fruit yield potential of a plant are approached, every management decision becomes increasingly important. Small errors can have a significant impact; therefore, every task needs to be performed without error in timing and process. Under such conditions, nutritional management of the entire growing system is absolutely essential.

Laboratory testing and diagnostic services are readily available in the United States and Canada as well as in other parts of the world. Samples can be quickly and easily sent to a laboratory from almost anywhere. Once the laboratory selection has been made, it is important to obtain from the laboratory its instructions for collecting and shipping samples before sending them. It is also important that the laboratory selected to do the analytical work is familiar with the type of samples being submitted and, when interpretations are to be made, whether the interpretation will be made by a skilled analyst.

With the analytical capabilities available today, together with the ease of quickly transporting samples and analysis results, growers can monitor their plant growing system on almost a real-time basis. Although a routine of periodic testing is time consuming and costly, the application of the results obtained can more than offset the costs in terms of a saved crop and superior quality production. The grower should get into the habit of routinely analyzing his water source, prepared and spent nutrient solutions, growing media, and crop plants. Interpretations and recommendations based on assay results are designed to assist the grower in order to avoid crop losses and product quality reductions.

Although laboratory analysis is recommended, on-site analysis is possible with the use of kits and relatively simple analytical devices. For example, elemental content determinations of solutions can be made on-site by the use of a HACH Chemical Company Water Analysis Kit. Although test kit procedures are available for determination of some of the micronutrients, laboratory analysis is recommended. However, concentration monitoring of the micronutrients is not as critical as monitoring of the major elements unless a micronutrient problem is suspected. For any diagnostic problem, laboratory analysis is always recommended, including all the essential elements—both the major elements and micronutrients.

Water analysis

The only way to determine what is in the water for irrigation and making a nutrient solution is to have it assayed. Knowing what is in the water will determine whether it is acceptable with or without treatment and whether adjustments would be required to compensate for constituents that are present (see p. 51).

Water available for irrigation or for making a nutrient solution may not be of sufficient quality (i.e., free from inorganic as well as organic substances) to be suitable for use. Pure water is not essential, but the degree of impurity needs to be determined. Even domestic water supplies, although safe for drinking, may not be suitable for some types of plant use. Water from surface groundwater sources, ponds, lakes, and rivers is particularly suspect, while collected rainwater and deep-well water are less so.

For the elements, the presence of Ca and Mg could be considered complementary because both elements are essential plant nutrient elements, whereas the presence of B and Na and the anions CO_3^{2-}, HCO_3^-, Cl^-, F^-, and S^- could be considered undesirable if levels are relatively high. The maximum concentrations of these elements and ions in irrigation water and water for making a nutrient solution have been established as presented in Chapter 4 (see pp. 51–52).

Testing for the presence of organic constituents is a decision that is based on expected presence. Surface waters may contain disease organisms and algae, while in agricultural areas, various residues from the use of herbicides or other pesticides may be in the water. Tomato, for example, is particularly sensitive to many types of organic chemicals; therefore, their presence in water could make its use undesirable, particularly for this crop.

Nutrient solution analysis

Since all the essential plant nutrient elements required by plants, except for carbon (C), hydrogen (H), and oxygen (O), are being supplied by means of a nutrient solution, errors in making the nutrient solution will affect plant growth, sometimes within a matter of a few days. The analysis of the nutrient solution should include pH and a determination of the concentration of the major elements—N (i.e., NO_3 and NH_4), P, K, Ca, and Mg—and micronutrients B, Cu, Fe, Mn, and Zn in solution. The concentrations of these elements can then be compared to what the designated formulation concentration is to be, and/or compared to what has been established as the optimal concentration range (see Table 4.11 in Chapter 4, p. 68).

Errors in the preparation of a nutrient solution due to weighing or volume measurements as well as the functioning of dosers (see pp. 76–79) are not uncommon—hence the requirement for an analysis to check on the final elemental concentrations prior to use. Since the elemental composition of the nutrient solution can be altered considerably in closed recirculating systems, it is equally important to monitor the composition of the solution as frequently as practical. A record of the analysis results should be kept and a track developed to determine how the concentration of the elements changes with each passage through the rooting media. On the basis of such analyses, change schedules, replenishment needs, and crop utilization patterns can be determined. The track establishes what adjustments in the composition of the nutrient solution are needed to compensate for the "crop effect"—not only for the current crop but also for future crops.

In addition, periodic analysis allows the grower to supplement the nutrient solution properly in order to maintain consistent elemental levels to ensure good plant growth as well as extend the useful life of the nutrient solution. Significant economy can be gained by extending the life of the nutrient solution in terms of both water and reagent use.

Water and nutrient solution in line monitoring

It is now possible to monitor the water and nutrient solution composition continuously with devices such as specific ion, pH, and conductivity meters, which are readily available at reasonable costs. The grower needs to determine how best to monitor water and a nutrient solution based on operating parameters and the requirements of the selected growing system.

Electrical conductivity (EC) is frequently used as a means of determining elemental replenishment schedules in closed recirculating nutrient solution growing systems (see p. 76). This technique is useful if previous knowledge is available as to which elements are likely to change and by how much. It is far more desirable to do an elemental analysis that quantifies each individual element and its ratio in the nutrient solution so that specific adjustments can be made to bring the nutrient solution back to its original composition.

Elemental analysis of the rooting medium

Elemental analysis of the rooting medium is an important part of the total evaluation of the elemental status of the rooting medium-crop system. When coupled with a plant analysis, it allows the grower to determine what elemental stresses exist in order to bring them under control. An analysis may be comprehensive, determining the concentration present in the growth medium by element, or more general, measuring the total soluble salt (EC measurement) content of effluent or by extraction of an equilibrium solution from the rooting medium. A comprehensive test is more valuable as a means of pinpointing possible elemental problems than just a determination of the EC of the effluent or extracted solution.

A test of an inorganic rooting medium, such as gravel, sand, perlite, or rockwool, measures the accumulation of salts that can significantly affect the elemental composition of the nutrient solution being circulated through it. Knowing what is accumulating in the rooting medium, it then becomes possible to alter the nutrient solution composition sufficiently to utilize the accumulated elements or to begin to calculate adjustments to the nutrient solution formula—with the idea of reducing the rate of accumulation while partially utilizing those elements already present in the rooting medium (see p. 92).

For those using perlite in bags or buckets or rockwool slabs, the recommendation is to draw an aliquot of solution periodically from the bag, bucket, or slab for assay. Based on either a complete analysis of this solution or only an EC determination, water leaching may be recommended to remove accumulated salts. In some management schemes, leaching of the rooting medium is performed on a regular basis as a matter of standardized routine. Systems following regularly scheduled leaching should also be subjected to periodic analysis of the growth medium effluent to confirm that the leaching schedule is effectively removing accumulated salts.

Plant analysis

The objective of a plant analysis (sometimes referred to as leaf or plant tissue analysis) is to monitor the elemental content of the plant in order to ensure that all of the essential elements are being supplied in sufficient quantity to satisfy the plant requirement as well as ensuring against elemental imbalances and excesses (Jones 2012a). The grower should develop a routine schedule of sampling and analysis during critical periods in the growth cycle of the plant.

Unfortunately, plant (leaf) analysis has largely been thought of as a diagnostic device, although its usefulness for monitoring is of greater significance. The procedure of routine sampling and analysis is frequently referred to as "tracking." Tracking provides the information needed to establish what nutrient solution management procedures are required to ensure that all of the essential element levels are within the sufficiency range for the crop plant being grown. It is well worth the time and expense to develop a track of elemental sufficiency in order to establish the proper nutrient solution management system firmly for future use.

The diagnostic role for plant (leaf) analysis is equally important. A grower faced with a suspected essential element deficiency or imbalance should verify the suspected insufficiency by means of plant (leaf) analysis. Many symptoms of elemental stress are quite similar and can fool the best trained grower or advisor. In addition, some stress conditions can be due to the relationship between or among the elements and therefore may require more than just a minor change in the nutrient solution formula to correct them. Without an analysis result, a change could be made that would only further aggravate the problem.

Since a plant (leaf) analysis requires the use of a competent laboratory, contact with the laboratory should be made before samples are collected and submitted. Most laboratories have specific sampling and submission procedures, which are important to follow. It is equally important to remember that different plant parts are not to be mixed together—such as leaves with stems or petioles, or selecting the whole plant as that sampled unless the plant is in its seedling stage of growth. Roots also should not be a part of a sample collected for analysis.

If no specific sampling procedures are given or known for a particular plant species, including the time for sampling, the rule of thumb is to "collect recently mature leaves below the growing point." Normally, the times for sampling are scheduled at major changes in the growth cycle, such as at flowering and initial fruit set. In addition, these same sampling procedures should be followed if the plant is being monitored periodically over the course of its life cycle, a procedure necessary to maintain a track of the elemental content.

For diagnostic testing, when visual symptoms of plant stress are evident, it is advisable to take similar plant tissues from both "affected" and "normal" plants. In this way, a comparison of elemental content can be made, which may be far more helpful in the interpretation than just an analysis of the stressed plant alone.

Great care should be used when selecting plants for sampling, as well as when selecting the plant part. In addition to what should be sampled, there are also avoidance criteria as to what not to sample or include in the sample:

1. Diseased, insect-damaged, or mechanically damaged plants or tissues
2. Dead plant tissue
3. Dusty or chemical-coated tissue

Tissue that is covered with dust or chemicals can be decontaminated by careful washing using the following procedure:

1. Prepare a 2% detergent solution and place in a large container.
2. Place the fresh leaf tissue in the detergent solution and gently rub with the fingers for no longer than 15 s.
3. Remove the tissue from the detergent solution and quickly rinse in a stream of flowing pure water.
4. Blot dry with a clean cloth or paper towel.

Great care is needed to ensure that the tissue being "washed" is not contaminated by some other substance present in the wash water or by contact with other substances or that the elements K and B are not being lost from the tissue in the washing process, as both can be easily leached if the time period for washing and rinsing with water is longer than that specified in the instructions.

Once the tissues have been collected, it is best to air dry them (one day in the open air is usually sufficient) before shipping to the laboratory for analysis. This will keep them from rotting while in transit, as any loss in dry weight will affect the analysis result.

The interpretation of an analysis result is performed by comparing the assay results obtained with established critical values or sufficiency range (Mills and Jones 1996). Interpretative values are applicable to plant tissue sampled and stage of plant growth when collected; therefore, it is important to follow the given sampling instructions so that the analytical results obtained can be properly interpreted.

General sampling procedures

When a diagnostic sample of water, nutrient solution, effluent from the rooting media, or plant tissue is collected for laboratory analysis, obtain the laboratory's recommended sampling (volume of solution required) and shipping procedures. Keeping the water and/or nutrient solution sample from being contaminated is essential; therefore, clean sampling devices and sample bottles or containers are to be used. For water or nutrient solution samples, one of the best sampling/shipping bottles is a new baby formula bottle. Remove the rubber nipple and tightly seal the lid after the sample has been drawn. When drawing a water sample or nutrient solution, run the water or nutrient solution for a few minutes, fill the bottle, dump, and then fill the bottle again.

The Internet

The Internet can be a significant source for obtaining diagnostic information. Today, it is possible for a grower to take a digital picture or video and send it to an expert for evaluation. Good photography skills are needed so that the photograph provides a good representation of what exists. The challenge for the grower is selecting the right individual(s) to make the evaluation and/or diagnosis, and then learning how to select information from the Internet that has foundation in fact and is reliable. As with a medical diagnosis, seeking a second (or even third) opinion is essential. Even the best experts can make misjudgments.

Outsourcing

With the increasing complexity of and many facets associated with the growing of plants hydroponically, proper management may be beyond the ability of any one individual. Therefore, the hydroponic grower needs to know to whom to turn when important decisions are to be made and/ or when a problem arises. Assistance may be provided by a well-trained and experienced county agent, crop consultant, or hydroponic supplier, but it is important that prior contact be made with such individuals to determine their degree of expertise so that time is not lost when a timely decision needs to be made.

Best Management Practices (BMPs)

Best Management Practices began with field crop production, although the basic principles have application to the hydroponic grower. The BMP manual written by the Potash & Phosphate Institute (Anon. 1991) defines how the diagnostic approach can be applied to any crop production system. Good Agricultural Practices (GAPs) established by the Food and Drug Administration and US Department of Agriculture (FDA/USDA) are "guidelines established to ensure a clean and safe working environment for all employees while eliminating the potential for contamination of food products." The trend is toward tighter regulation of chemical use that will equally apply to the hydroponic grower.

Appendix D: Common errors made when plants are grown hydroponically

Making wise, timely decisions and avoiding errors are essential for success when growing plants hydroponically. Having a basic working knowledge of the hydroponic method as well as how plants grow is essential. In this appendix some of the common errors that I have observed and experienced are described, as well as those factors I have found that lead to successful growing. These factors are arranged by category.

Nutrient solution formulations

1. The use of so-called "pure water" for formulating a nutrient solution or for irrigation use is not necessary for most hydroponic growing systems, unless the source water contains fairly high concentrations of potentially damaging elements (see p. 50).
2. Most hydroponic nutrient solution formulations are more concentrated in elemental content than needed and frequently lack the proper balance among the elements, particularly the major elements—K, Ca, and Mg—in solution (see p. 85).
3. Phosphorus (P) concentration excess is probably the most frequently occurring insufficiency in most nutrient solution formulations, while Mg and Zn are the elements most frequently inadequate in concentration (see p.83).
4. Combined with the hydroponic growing method, the use factors, volume applied with each irrigation, and frequency of irrigations associated with a nutrient solution formulation are as important as the elemental content of a nutrient solution formulation.

5. With increasing volume of nutrient solution applied with each irrigation and increasing number of irrigations, the more dilute the nutrient solution formulation should be.

6. For the majority of growers, the most economical procedure is to make their own nutrient solution formulation (see p. 55 for list of required reagents), rather than selecting from the many prepared formulations since a majority of these formulations are not well suited for the commonly used hydroponic growing systems and selected crop plants.

7. The inclusion of an organic substance, such as humic acid, or other similar substances or an array of microorganisms into a nutrient solution will not benefit plant growth and has the possibility of an adverse effect. In addition, the inclusion of an organic substance in a nutrient solution is an invitation for root disease invasion.

8. Recirculation (reuse) of a nutrient solution may require treatment, such as restoration to its initial volume by adding water, adjusting the pH and elemental content, removal of suspended material by filtering, and sterilization to inactivate disease organisms.

9. A well formulated nutrient solution can be used for most plant species at various stages of plant growth and for most hydroponic growing systems—therefore not requiring specific formulations for all these varying conditions.

10. The inclusion of nonessential plant nutrient elements in a nutrient solution formulation is not needed; the only exception may be for the element silicon (Si) (see p. 42).

11. After making a nutrient solution formulation, the nutrient solution should be analyzed to determine its element content concentrations. If the nutrient solution is being formulated using injector pumps, such an analysis is essential to ensure that the injector pumps are operating properly and the nutrient element concentrates are at their proper elemental concentration—either based on the formulation or compared to the desired range in concentration for optimum plant growth (see Table 4.11 in Chapter 4, p. 68).

Hydroponic growing systems

1. Not all hydroponic growing systems are well suited for a particular use and/or plant species. A good example is the NFT growing system, which is not suitable for use with long-term crops such as tomato, cucumber, and pepper, as the root mass will fill the NFT trough and impede the flow of nutrient solution down the trough.

The root mass becomes anaerobic, resulting in the death of some roots, which results in a reduction in plant growth or even possible death of the growing plant.

2. Most hydroponic growing procedural recommendations are wasteful in their use of water and reagents. Therefore, application procedures need to be designed in order to obtain maximum utilization of applied irrigation water and the nutrient solution. This will require careful monitoring of water and nutrient solution use coupled with experimentation in terms of adjusting timing and quantity applied based on plant requirements for water and/or nutrient elements.

Rooting media

1. The physical and chemical characteristics of a rooting medium may affect the nutritional status of the growing plant—either by contributing to the nutrient element requirement of the plant or by participating in the interactions that may occur between the rooting medium and the applied nutrient solution (see pp. 83–84).
2. Some of the commonly used rooting media may contain sufficient quantities of a plant essential element so that element does not need to be included in a nutrient solution formulation. The elements that may be sufficiently supplied by the rooting medium are K, P, Mg, Cu, Mn, and Fe (see pp. 93–97).
3. With the drip irrigation hydroponic growing method, there occurs an accumulation of elements in solution and as precipitates. The common procedure is to monitor the EC of nutrient solution being discharged from the rooting medium or from an aliquot of solution being retained in the rooting medium. When the EC exceeds that of the applied nutrient solution, the rooting medium is to be leached with water. This accumulation of what is known as "salts" can be minimized by reducing the elemental content of the nutrient solution and/or by alternating between a nutrient solution application and water only. The accumulation of "salts" in the rooting medium is an indication of poor management of the use of a nutrient solution formulation.
4. Accumulation of elements in solution as well as precipitates can also occur in the rooting medium with the flood-and-drain method and in the root mass for plants being grown using the NFT method.

Plant growth and maturity

1. Insufficiency of an essential plant nutrient element may not appear as a visual symptom, although plant vegetative growth and product production (flowers and fruit) will be less than the potential. Monitoring the nutrient element content of the growing plant can be used to determine nutrient element sufficiency (see p. 86).

Grower skills

1. Although skill is needed on the part of a grower to manage a hydroponic growing system efficiently, there are also considerable "commonsense" or "green thumb" aspects that can contribute to a grower's success, even though the grower may be carefully following operating instructions (see p. 123).
2. Past experience can contribute to the ability of a hydroponic grower to grow plants at their genetic potential in terms of both vegetative growth and product yield.

Disease and insect control

1. A commonly occurring error is not to be prepared to deal with the occurrence of an insect infestation or disease occurrence as well as not seeking professional assistance for identification and recommendations for best control treatments. Avoidance is the best control measure, followed by knowing when to apply an effective control treatment.

Miscellaneous

1. Most plants can be grown hydroponically fairly easily, except for the root crops, such as potato, radish, beets, etc., although they have been successfully grown using uniquely designed hydroponic systems.
2. Most of the failures associated with hydroponic growing are due to either the infestation of root disease or the inability to control the nutrient element environment of the rooting media that results in plant nutrient element insufficiencies. This sometimes results in easily identifiable visual symptoms, as well as not visually seen but reducing plant vegetative growth and low crop product yield and quality.
3. The hydroponic knowledge base is very large with far more misinformation available than that which is true. The challenge for the hydroponic grower is to be able to separate fact from fiction.
4. Growing plants hydroponically is not a means for overcoming those growing conditions that will reduce plant growth, such as improper air and root temperatures, low or high light intensity, light spectrum

and duration factors, low or high (resulting in wilting) atmospheric demand, and stagnant air or high air movement over plant leaf surfaces.

5. Failure to take timely preemptive measures as well as not anticipating changing conditions that will impact plant growth can result in poor plant growth and product yield irrespective of the hydroponic growing system used.

6. Poor performance from the hydroponic growing system employed can be due to the failure of a grower to seek assistance from those with professional skills and experience regarding everyday operating procedures as well as when faced with unexpected poor plant growth or appearance.

Index

Printed in the United States
by Baker & Taylor Publisher Services